AF701053

Electrifying India

Awarded the

Joseph W. Elder Prize
in the Indian Social Sciences

by the American Institute of Indian Studies and published with the Institute's generous support.

AIIS Publication Committee:

Susan S. Wadley, Co-Chair
Brian A. Hatcher, Co-Chair
Joyce B. Flueckiger
Pika Ghosh
Priti Ramamurthy
Steven Wilkinson

Electrifying India

REGIONAL POLITICAL ECONOMIES OF DEVELOPMENT

Sunila S. Kale

Stanford University Press
Stanford, California

Stanford University Press
Stanford, California

©2014 by the Board of Trustees of the Leland Stanford Junior University.
All rights reserved.

No part of this book may be reproduced or transmitted in any form or by any means, electronic or mechanical, including photocopying and recording, or in any information storage or retrieval system without the prior written permission of Stanford University Press.

Printed in the United States of America on acid-free, archival-quality paper

Library of Congress Cataloging-in-Publication Data

Kale, Sunila S., author.
Electrifying India : regional political economies of development / Sunila S. Kale.
pages cm
Includes bibliographical references and index.
ISBN 978-0-8047-8796-3 (cloth : alk. paper)
1. Rural electrification—India—History. 2. Electric utilities—Government policy—India—History. 3. Rural development—Government policy—India—History. 4. India—Economic policy—1947– I. Title.
HD9688.I52K35 2014
333.793'20954091734—dc23 2013045174

ISBN 978-0-8047-9102-1 (electronic)

Typeset by Westchester Publishing Services in 10.5/13.5 Bembo

In memory of my father
Sharatkumar Vishnu Kale (1940–2013)

Contents

Maps and Tables

Maps

Tables

Abbreviations

APERC	Andhra Pradesh Electricity Regulatory Commission
APGENCO	Andhra Pradesh Power Generation Corporation
APSEB	Andhra Pradesh State Electricity Board
BEST	Bombay Electric Supply and Transport Company
BJP	Bharatiya Janata Party
BSES	Bombay Suburban Electric Supply
BSP	Bahujan Samaj Party
CAD	Constituent Assembly Debates
CEA	Central Electricity Authority
CERC	Central Electricity Regulatory Commission
CESC	Calcutta Electric Supply Company
CPI(M)	Communist Party of India (Marxist)
Cr.	crore (equal to 10 million rupees)
DMK	Dravida Munnetra Kazhagam
DPC	Dabhol Power Company
DVA	Damodar Valley Authority
GEOLRO	State Commission for the Electrification of Russia

GP	Ganatantra Parishad
GRIDCO	Grid Corporation of Odisha
hp	horsepower
IPPs	Independent Power Producers
IPS	irrigation pump set
JD	Janata Dal
kW	kilowatt
kWh	kilowatt-hour
LNG	liquefied natural gas
MEDA	Maharashtra Energy Development Agency
MERC	Maharashtra Electricity Regulatory Commission
MLA	member of the Legislative Assembly
MNES	Ministry of Non-Conventional Energy Sources
MNRE	Ministry of New and Renewable Energy
MOU	Memorandum of Understanding
MSEB	Maharashtra State Electricity Board
MW	megawatt
NHPC	National Hydroelectric Power Corporation
NPTC	National Power Transmission Corporation
NSDP	net state domestic product
NTPC	National Thermal Power Corporation
OERC	Odisha Electricity Regulatory Commission
OPGC	Odisha Power Generating Corporation
OSEB	Odisha (formerly Orissa) State Electricity Board
PPA	power purchase agreement
REC	Rural Electrification Corporation
Rs.	rupees
SC	Scheduled Caste
SEB	State Electricity Board

SERC	State Electricity Regulatory Commission
ST	Scheduled Tribe
T&D	transmission and distribution
TDP	Telugu Desam Party
TVA	Tennessee Valley Authority
UNDP	United Nations Development Program
USAID	United States Agency for International Development

Acknowledgments

I started thinking about the research for this book in 2000. From then to now, I've collected an array of debts to friends, colleagues, family, and institutions, which I acknowledge with pleasure.

I am grateful to many institutions and interlocutors in India. A large number of individuals at electric utilities in Mumbai, Delhi, Hyderabad, and Bhubaneswar shared their insights, patiently answered my questions, and gave me access to valuable data. At Pune University, conversations with the late Rajendra Vora helped me to see how my research on electricity offered a novel view of Maharashtra's political economy, while Suhas Palshikar generously commented on portions of the manuscript. The team at Prayas Energy in Pune provided an astute audience when I shared my research in 2013. The librarians and staff at the Gokhale Institute of Politics and Economics located a number of materials and older government documents, and allowed me to use the cool basement reading room to complete revisions in 2013. In Delhi, Navroz Dubash at the Centre for Policy Research supported the project with helpful discussions at the early stages and gave valuable comments on later portions of the manuscript. In Bhubaneswar, I was fortunate to meet a group of committed intellectuals and activists who illuminated many aspects of Odisha's politics and sociology for me. I am indebted to Anita Mishra, Banikanta Mishra, and Birendra Nayak. In my efforts to locate archival materials in Odisha, Bani was an enthusiastic and generous guide and, more recently, both he and Dr. Nayak read and commented on portions of the manuscript.

Cathy Boone has been a model scholar and mentor, intellectually generous and encouraging. Sumit Ganguly has shown unflagging support over the years for which I am very thankful. I am also grateful to Jamie Galbraith, Clement Henry, and Raul Madrid for seeing me to the finish line at the University of Texas. I have benefited from conversations with and feedback from of a number of other colleagues and friends about this project and others. My thanks to Bidisha Biswas, Francine Frankel, John Harriss, Devesh Kapur, Jason Kirk, Atul Kohli, Nimah Mazaheri, Rahul Mukherji, Karthika Sasikumar, Aseema Sinha, and Ashutosh Varshney. For their friendship and collegiality in Austin, I'm grateful to Oksan Bayulgen, Jon Bloch, Neil DeVotta, Jeff Ladewig, and Brian Wampler. Also, thank you to anonymous reviewers, whose incisive comments have undoubtedly strengthened this book.

At the University of Washington, I have found a vibrant and supportive intellectual home. My sincere thanks to Nidhi Agrawal, Jameel Ahmad, Manish Chalana, Dan Chirot, Frank Conlon, Sara Curran, Jennifer Dubrow, Kathie Friedman, Don Hellmann, Judy Howard, Reşat Kasaba, Sonal Khullar, Wolfram Latsch, Tony Lucero, Sudhir Mahadevan, Saadia Pekkanen, Cabeiri Robinson, Michael Shapiro, Matt Sparke, Jim Wellman, and Anand Yang. Sareeta Amrute, Paul Brass, Maria Elena Garcia, Joel Migdal, Robert Pekkanen, Scott Radnitz, and Priti Ramamurthy shared valuable comments, for which I am very appreciative.

For supporting language training and research, I thank the American Institute of Indian Studies, the Government Department at the University of Texas, and the South Asia and Global Studies Centers at the University of Washington. A fellowship at the Simpson Center for the Humanities in 2011–12 enabled me to write much of the manuscript, and I thank Kathleen Woodward and colleagues in the Society of Scholars for an intellectually stimulating year. At Stanford University Press, Geoffrey Burn has shown faith in this project and expertly guided me through the editorial process, for which I am grateful. For preparing the manuscript, my thanks to James Holt, Frances Lyon, Bill Nelson, and Rachel Lyon.

I am exceptionally lucky in my friends, many of whom have seen this project from the start. For their fellowship over the years, I am grateful to Paige Bartels, Gayatri Chatterjee, Jae Chung, Leah Clarkson, Suzanne Cox, Whitney Cox, Rana Dasgupta, Naisargi Dave, William Elison, Crispin Faber, Jürg Faber, Catherine Kamerling, Henry Kamerling, Renee Kaufman, Elizabeth Kolsky, Srinivas Krishnan, Surabhi Kukke, Nicole Mellow, Lisa Mitchell, Sucheta Paranjpe, Ram Rawat, Uzma Rizvi, Andy Rotman, Svati Shah, James Smartt, and Murtaza Vali.

Family in India and the United States have buoyed me and this project with grace and patience, even though it was not always clear to them what I was doing and why it went on for so long. The Novetzkes and Greenes have welcomed me into their families with much warmth. To the Paranjapes, Baiskars, Rasthes, Gogtes, and Phataks in Thane and Mumbai, and to the Kales and Leles in Pune, my thanks for giving me so many loving homes in India. My grandparents, Aii and Tatya, were pleased and proud when I completed my PhD, and I know that Tatya would have been just as happy to see this book. I thank my sisters, Vidula and Minal, and their delightful families for their enduring affection and humor. I find it difficult to adequately express my gratitude to my parents for their persistent love and encouragement. They are my models for how to live life with kindness and creativity. Sahil and Siyona, your irrepressible joy, conversation, and curiosity have made life incomparably more delicious than it would have been. And finally, Christian, you have been an ardent ally for fifteen years; I know how lucky I am.

Electrifying India

1

Introduction

> Electricity is perhaps the most necessary and the most revolutionary thing which you can take into the rural areas. The moment you take electricity, all kinds of things begin to move.
>
> Petty industries grow up, agriculture is affected; everything is in fact affected.
>
> The whole life of the people is changed.
>
> —*Jawaharlal Nehru, prime minister of India, 1947–1964*

Since electricity's invention in the late nineteenth century, the spread of electric utilities has come to signify the advance of modernity. The representation of electricity in India is no different. Consider for instance its role in *Swades*: *We, the People* (*Homeland: We, the People*),[1] a 2004 Hindi film that received critical acclaim in India and abroad. *Swades* tells the story of Mohan, a US-based NASA employee who returns to India in search of his childhood nanny. Mohan, played by Bollywood star Shah Rukh Khan, finds her living in a village that is plagued by a host of problems commonly ascribed to rural India: poverty, social discrimination, and a general resistance to ideas of progress. A few turns of the plot and a rising love interest convince Mohan to dedicate his time in India to improving conditions in the village. With his money, technical knowledge, and a few local recruits, he builds a small hydroelectric plant on a hillside just outside the village. Watching as electricity wires are strung along improvised poles, the older women of the village express awe that *bijli* (electricity) is finally coming to them. Mohan gives the signal to release the flow of water through narrow channels, and as the waves hit the turbine, everything changes, both for Mohan and the villagers. The khadi-clad local politicians, who had doubted the project from the start and had contributed neither their own labor nor any of the state's development funds, stand aside as electricity travels through the makeshift grid to light a single bulb in the hut of one of the village's oldest residents. The villagers cheer and embrace the promise of technological progress. Mohan finds romantic as well as patriotic love, leaving his coveted foreign post to resettle in India. We end with Mohan, having devoted himself to developing India through electrification, now himself Indianized as a result of his efforts to modernize the village.

The film's hero inherits a belief in the capacity of electricity to not only improve material conditions but also catalyze broader social change, and this belief has been a mainstay of the Indian nationalist vision for modern development, and indeed for nationalist modernizers everywhere throughout the twentieth century. When Nehru spoke the words that open this chapter, more than three-quarters of India's population lived in rural areas, and only a small fraction of the total population had access to electricity. Nehru's belief in the power of electricity was a belief in the potential transformation of India in the postindependence era as well as a vision of its unification through modernization. Electricity would make factories hum, irrigate farms, illuminate public spaces, and lengthen the study day of Indian students. Practically, the network would consist of generating plants that could take advantage of India's plentiful coal supplies and ample rivers, and a robust transmission grid to transport electricity over long distances and through dense distribution networks that would penetrate cities, towns, and even the most remote Indian villages. The aim was to electrify India as a whole, bring it fully into the developed world, and in doing so, unify a socially and politically diverse country.

Nehru's ambition has not been realized yet. Despite four decades of planned economic development followed by close to twenty years of dramatic economic growth and despite the fact that electricity is now considered to be essential, today more than 400 million mostly rural Indians have no access to electricity. India, with 17 percent of the world's population, accounts for close to 40 percent of the world's population without electricity. There are thousands of real examples of the fictional village in *Swades*, which was itself based on the story of diasporic Indians who returned to their homeland to create sustainable rural electrification projects.

This book begins by asking why, despite the intentions of national politicians like Nehru and the commitment of development planners, the central state in New Delhi has not exercised its writ across India's heterogeneous federation to make electricity, a quintessential commodity of the twentieth century, available to all Indians. Today there is tremendous variation in electricity access across states; in some, nearly all households have electricity, while in others, it is available to fewer than one in five (see Map 1.1).

The political and social character of India's subnational units vary widely, one consequence of which is that the state looks and acts differently in dif-

MAP 1.1 *Household electricity use for lighting in India, 2011*
SOURCE: *Census of India, 2011.*

ferent places. In *Electrifying India*, I argue that it is this variation in the social and political foundations of the state at the provincial level that accounts for variation in patterns of infrastructural development, particularly in the crucial decades from the 1960s to the 1980s. In those parts of the countryside that were successfully electrified, the gains were due to neither nationalist idealism nor only technocratic plans. Instead, rural electrification occurred either when rural constituencies became politically influential in state governments or when farmers mobilized to demand a larger share of development resources. The initial conduit of electricity into rural India was for its productive impact in agro-industries and for irrigation; household access followed. The results from rural electrification have often been mixed. Certainly, where it occurred, rural electrification profoundly impacted agrarian productivity as well as rural standards of living. But it also has led to ecological crises, eroded the stability of the electric grid, and in some cases, exacerbated inequality between rural citizens.

In this book, I explore a global phenomenon of modernization and development. Building electric infrastructure was one of the signal pursuits of modernizing polities throughout the twentieth century, and India was no exception. Just as railroads were fundamental to the project of state building during the British colonial period, and emblematic of the "positive" effects of a modernizing colonial force, an extended and interconnected electricity grid was at the heart of postcolonial India's development plan as well as its creation as a unified modern nation. From Nehru's pursuit of big dams to generate hydroelectric power to his early patronage of a civilian nuclear energy program, electrifying India has been the backbone of the nation's transformation during the last sixty-five years. As an indication of energy's centrality, consider that during the first thirty years after independence, despite myriad pressures on public finances, nearly one-fifth of all planned expenditure was devoted to the power sector, making it among the largest categories of public spending.

A compelling body of work has shown that the process of electrification, while highly technical, is never neutral. In every instance, social and political contexts shape the way that electricity becomes embedded in a given place. In turn, the modalities of electrification—whether state owned, privately owned, or mixed sector; whether highly local or highly centralized; whether produced by water or fossil fuels—determine what kind of impact electricity will have.[2] For example, in the mostly state-owned electric systems of Scandinavia, the Netherlands, and Germany, most homes were electrified by 1930. By contrast, levels of household electrification in America, where private utilities predominated, were far lower in the same period. Instead, American cities benefited from more public lighting and commercial advertising, spectacular displays that were much more rare in Europe at the time.[3]

In India, too, the process of electrification has been conditioned by social and political contexts that vary from state to state. I explore these processes and contexts in Chapters 2–5. My analysis is bracketed by two pieces of legislation: the Electricity (Supply) Act, 1948, and the Electricity Act, 2003. The first established the massive state-owned utilities—State Electricity Boards, or SEBs—which the second dismantled, clearing space for private firms, independent regulation, and market competition. Chapter 2 examines debates and conflicts throughout this period from the vantage of New Delhi. In the subsequent chapters, I turn to three Indian states—Maharashtra, Odisha, and Andhra Pradesh—to explore the conditions and consequences of electrification in each, from the time of their formation until 2003.[4] Together the four chapters show how a single set of rules emanating from Delhi was implemented with divergent effects in federal India.

In exploring the politics and history of electrification in India, this book also thinks through two sets of larger debates. The first explores the political economy of market reforms. I probe the longer politico-historical connections across two time periods that are usually treated as distinct: the period of state oversight and ownership from the 1950s to the 1980s, and the period of market reforms from roughly the early 1990s onward. I argue that the earlier period of an advancing state apparatus conditioned in important ways the manner of the state's retreat in the following period. My aim here is to historicize market reforms across these two political-economic periods in India.

A related concern is to understand the role of the Indian state in development. Despite the inordinate authority of the central Indian state to determine the country's development agenda, collect and disperse revenue, and wield military power, India's states have commanded authority in vital domains, including key pieces of the social and physical architecture of the Indian state and economy. I demonstrate this through the empirical lens of electricity, which I suggest is a quintessential piece of the modern infrastructural state as well as a critical feature of the national imagination of Indian development and modernity. State-level regimes produced and allocated goods in sectors like electricity in ways that attempted to advance distinct political economic visions, albeit sometimes unsuccessfully. These political economic ideologies were the product of the character of the regional state and its interactions with regional constituencies, both rural and industrial. In no case was electricity simply provided; where electricity appeared, a politics of development and differentiation was at play.

In the next section, I position my book by briefly surveying the theorists and debates that have most influenced my thinking in this project. I do not expound at length on these literatures in order to give more space to the empirical substance of the book. However, these bodies of scholarship have been instrumental in shaping my understanding of the politics of market reforms and the role of the Indian state in development.

Historicizing Market Reforms

Why and how do nation-states vacate economic spaces to the forces of competition and private ownership? This question has inspired a voluminous literature on the global turn to the market, with books and articles published almost continuously on the topic from the early 1980s to the present. In India, it is increasingly provincial governments and not the central government that dictate the speed and direction of market reforms. Examining the process of change in the electricity sector—the purview of the state, not central, governments—then, is a useful entrée to the dynamics at work in the broader arena of India's market reforms. And yet here, many of the dominant insights about market reforms fail to fully make sense of why electricity restructuring and privatization in India has been so uneven across states.

When have market reforms occurred and who has pushed them forward? Multiple explanations have been offered in recent years—none of which comes to grips with the uneven process of privatization and restructuring in the electricity sector. Even with key politicians and technocrats backing privatization, some states have stumbled badly in effecting reforms, while others have unaccountably succeeded. Much of the work on the political economy of market reforms has focused on prevailing institutions, coalitions, and ideologies to explain the timing and manner of market reforms. Many of the theoretical differences within this body of literature pivot around how both the state and state-society interactions are conceptualized, and whether economic policies are believed to be driven by social interests, the initiatives of political elites, or some feature of their interactions.

Among those who emphasize the importance of social forces and state-society interactions, much of the debate revolves around understanding how winners and losers of reforms align themselves politically vis-à-vis a market-reforming state, yielding several sets of contrary insights. On the one hand, we can expect that when groups in society experience the inevitable costs of reforms, the losers will push back and force once-radical measures to be tempered and hence made less effective.[5] On the other hand, we know that market reforms too have distributional implications, and coalitions of actors who stand to benefit will push them forward if possible.[6] A focus on social alignments is useful to understand the dynamics at play in the electricity sector, the distributional implications of which are often loud and clear. Without an explicitly historical analysis, though, we are left taking social alignments as a given rather than understanding how they are produced by previous eras of state intervention.

Shifting the focus from social interests and their interactions with the state to the state itself more directly, a body of literature specifies which types of state elites, operating under what kinds of conditions, are likely to drive market reform. Some scholars emphasize that political outsiders, whose tenure is not associated with the origins of existing institutions, act with greater decisiveness.[7] Related to this is the idea of a "change team," made up of a set of willing and capable technocrats who are politically isolated but have the crucial backing of the head of state.[8] To further highlight the importance of political elites and their cognitive frames, some scholars suggest that an ideological sea change is an essential prerequisite for new policy initiatives.[9] Running through many of these accounts is the importance of economic crisis as a proximate cause of the market reforms of the 1980s onward. Such a crisis clears space for a new wave of political actors to enter the scene and dismantle the existing institutional architecture.[10] A less political version builds on the idea that technocratic rationality can guide political and policy change in the aftermath of economic crises.

Many of these theories of market reforms have illuminated aspects of the Indian experience.[11] And yet the unevenness in market reforms in the electricity sector remains unexplained even by this substantial tool kit of causal arguments. In electricity, because market reforms threaten to take away valuable subsidies from politically influential constituencies, reform by "stealth" is not as viable as in other sectors.[12] The most committed advocate of reforms in India during the 1990s at the state level was arguably Andhra Pradesh's chief minister, N. Chandrababu Naidu. Although he commanded a great deal of respect from international financial institutions as well as like-minded policy makers in New Delhi, his plans to restructure and privatize his state's public utility fell short against popular protests and opposition. Likewise, a purely technoeconomic reckoning of the electric utility industry cannot help us to explain which state governments chose market strategies and which rejected them. Although the first state to embrace utility privatization—Odisha—did not have the strongest utility, neither was it the worst off.[13] Although many existing accounts of that state's privatization emphasize the financial straits of the state government, my research finds this fact to be important but not determinative.

Despite the obvious as well as subtler differences across this scholarship on the political economy of market reforms, one feature that much of it shares is a temporal focus on the years, months, and electoral cycles that just precede market reforms. My proposition in this book is that how and with what effect states intervened in the economy in a much earlier period of state expansion will add to our understanding about whether and how states retreat in the contemporary period. Such a perspective can contribute to explaining the substantial variation that persists in how states have gone about turning to the market, despite the pressures of policy convergence in a globalizing world. My aim is to put market reforms into longer genealogies of Indian states' variable engagement with economic sectors and spaces.

To get traction on this question of how the past constrains and informs the present, I analyze three constitutive units of federal India: Maharashtra, Odisha, and Andhra Pradesh. For the past two and a half decades, India has been, like so many other countries around the world, in the midst of what is variously described as a neoliberal economic reorientation, a market transition, or a deepening integration with a globalized world economy. By looking at one sector of the economy—electricity—I can isolate more clearly the variables and mechanisms conditioning state retreat in the era of economic reforms. Specifically, I argue that the differential manner of state intervention in the electricity sector from the 1960s through the 1980s configured subsequent state retreat. Rather than viewing utility privatization and restructuring as a function of political will or ideological reform-mindedness, I show that precisely those subnational political regimes that had used the tool of electricity quite successfully to transform agricultural production from the 1960s to the 1980s were the least likely to carry out market reforms in the 1990s. Instead of pursuing the model of restructuring and privatization favored by both the World Bank and the India central government, these states opted to maintain the basic structure of their utilities while finding alternative ways to augment electricity supply and utility profitability. By contrast, subnational units with the weakest record of rural infrastructure development in the earlier decades were paradoxically best positioned to pursue market reforms in the 1990s.

Much of the literature on market reforms has not focused on the rural sector. Analyzing the electricity sector has the further benefit of allowing us to see how the rural sector has responded to market reforms, and occasionally the might of the countryside. The story of Andhra Pradesh's failed electricity privatization is perhaps the best illustration of this. There, protests from farmers and from politicians who championed their cause were enough to stall privatization.

Electricity and Infrastructural Power

There are several characteristics of electric utilities that make this sector a particularly useful tool with which to understand political economic logics, and it is here that this book engages the debate about the role of the Indian state in development. First, electrification has vast material consequences. Electricity can dramatically transform all kinds of productive activity. We can think of the switch from hand looms to power looms in the textile sector, from well water drawn by animal or human labor to groundwater accessed by an electric pump on a tube well, or of techniques of industrial production and information technology that are nearly impossible in its absence. Electricity also transforms the ways human live outside of the formal realm of "work," from the ways we organize lives in private to the character of public space. A well-lit nighttime street, whether in a village or a city, affects public safety and provides novel opportunities for sociability and commerce alike. One further dimension of electricity and twentieth-century projects of electrification is their symbolic association with modernization, as an illuminated sign of modernity. As many Indian nationalists and political leaders understood, this gave electricity a symbolic or discursive importance in addition to its physical and material value. I take up this aspect of electrification in Chapter 2.

Second, electricity and electrification have political valence, constituting one facet of twentieth-century infrastructural power. Scholars have noted that changes in technology help to explain why and how processes of state building have varied over time. If military technologies governed the success and longevity of early states, advances in bureaucratic technology extended both the time horizon and the spatial reach of more complex modern states. By the nineteenth century, public works became an increasingly important means of stabilizing rule. As one historian of South Asia notes, by the mid-nineteenth century, "direct state investment, especially in irrigation . . . was understood then as being complementary to military pacification in the strengthening of empire."[14] Along with massive investments in canal irrigation throughout the Indian subcontinent, railroad financing and construction had a similarly salutary effect on imperial ambitions to command. Railroads both enhanced revenue by facilitating trade and allowed the imperial state to make its way on the ground more quickly to quell protest.

For twentieth-century states, the equivalent technology to railroads was electricity. In the late nineteenth and early twentieth centuries, electricity systems in much of the world were concentrated in cities and privately owned. By the time of Indian independence in 1947, though, the political and economic importance of electricity was such that in most newly independent countries, electrification was taken on as a state project, something too vital to be left to the vicissitudes of the private profit motive alone.

I take "infrastructural state" from Mann's twofold delineation of the "autonomous power" of the state, or that aspect of state authority that is not dependent on structures or actors in civil society.[15] The first aspect, "despotic power," is the more traditional meaning of state power. The state's monopoly on legitimate violence allows it to punish and reward without need to negotiate or compromise with other actors or groups in society. Mann identifies the second face of state power as "infrastructural power," by which he means "the capacity of the state actually to penetrate civil society, and to implement logistically political decisions throughout the realm."[16] Modern liberal democratic countries are examples of states with low despotic power (at least relative to autocracies or monarchies) but increasingly capacious infrastructural power. Four mechanisms facilitate the expansion of infrastructural power: increasing specialization of state activity; literacy, which enables consistent communication of the state's message, including its legal order; measurement to facilitate exchange; and the expansion of technologies to facilitate the movement of people, goods, and ideas.

For modern states, electricity is both object of and mechanism for the expansion of infrastructural power. Electricity enables the state's message to fill the airwaves, facilitates transportation, and has an indirect effect on the spread of education through improved schooling and studying. Networked energy in particular denotes in bald terms the spatial reach of the state, showing where the state has marked its territories through electricity poles and transmission lines. The flip side of this is also true: the geography of the energy grid makes evident where the state exists only thinly on the ground, corresponding to "brown areas" that show a state's incapacity to govern uniformly.[17]

Mann's idea of the state's infrastructural power intersects and partially overlaps with another line of inquiry, which links modern states with the project of development. For India—like many other countries that claimed sovereignty in the mid-twentieth century and after—providing "development" as measured by certain economic indicators of progress and growth constituted one of the main sources of state legitimacy. India's minister for works, mines, and power, N. V. Gadgil, voiced this sentiment while debating the Indian Constitution, pronouncing that "a modern state is something more than a police state; the modern state has come to mean a social service state."[18] To be sure, the despotic state continued to mark its presence; particularly when faced with external or internal aggression, the Indian state has been quick to deploy its military and police forces to shore up its sovereignty. But rhetorical commitments, public expenditures, and bureaucratic labors suggest that for the Indian state, the goal of development was the more prominent of the state's two faces. The common understanding of what it means for a state to undertake development maps almost cleanly onto the mechanisms for the expansion of the infrastructural state: increase literacy, build roads, extend communications networks, and electrify. A state with infrastructural power, then, is a developed state.

The ideal typical state of Mann's original formulation was a centralized one. How do federal arrangements complicate the notion of infrastructural power where authority is truly shared and regularly contested between center and region/state? Mann, along with several other scholars, briefly visited this idea in a recent volume of essays devoted to the concept of the infrastructural state.[19] One essay analyzes variation in public health systems in early twentieth-century German cities to delineate a "'bottom up' path to infrastructural power."[20] Rather than explain public goods provision by looking at social preferences, which is a common tendency, the author examines the interaction between social preference and institutional capacity, not at the national level but at the subnational level. Doing so also helps resolve the paradox of a regime—imperial Germany—that is considered to be a weak infrastructural state yet had robust local institutional capacity and robust public goods provision. One lesson to take from this is that how we measure infrastructural strength needs to be attuned to the multiple strata of political authority.

Despite this proposition that infrastructural power migrates not only from the top down or from the center outward, but perhaps emanates from several nodes, the approach of many is still to assume that infrastructural strength is an attribute of the central state.[21] Of the political fragmentation in federal polities like Germany and the United States, Mann suggests that increasing coordination across the national space ensures that in the end, "enduring federalism . . . adds complexity to our discussions of infrastructural power, but they produce only mild subnational variation."[22] But as Mann suggests later in the same essay, infrastructural power has not developed in the same way in postcolonial states, a proposition that may be particularly true of federal polities. In the case of India, I argue that the process of infrastructural state building—that is, the coordination of those very mechanisms like electricity, communications, and education that are reflections of the infrastructural power of the state—was in many important ways a decentralized one.[23] Taking Mann's formulation of infrastructural stateness as a starting point leads to a modification of the axiomatic view of Indian federalism.

The Infrastructural State and the Unique Problems of Indian Federalism

In most of the literature on federalism in general and on Indian federalism in particular, India exemplifies a version of federalism in which authority is heavily tilted toward the center.[24] A recent analysis of the design of federal institutions lists India as only a quasi federation because it has only two of the three characteristics of full federations.[25] And indeed, judging by how the functional domains of authority are allocated between center and state, on balance, the Indian Constitution's formal distribution of powers favors the central government. The most often cited evidence of India's heavily centralized federalism are the powers of emergency (which allow the central executive to dismiss state governments) and the prerogative of the central government to redraw subnational boundaries. A numerical accounting of the parceling of legislative arenas between center and states provides additional support for this view.[26]

In a recent article, Rudolph and Rudolph reanalyze the historical antecedents of contemporary Indian federalism to argue that federalism in India is neither an incidental nor an unimportant feature of Indian state power.[27] Rather, by showing how federalism is constitutive of state formation, the Indian story helps to challenge normative accounts that privilege monolithic sovereignty. After providing a counternormative account of state formation that emphasizes federalism, they end by agreeing that the federal equation won by Nehru and others at great cost tilted "decisively toward a strong center."[28] As they argue in a separate essay, a return to a more decentralized state would await the era of market reforms.[29]

Scholarship on the foundational period when the constitution was drafted, from 1948 to 1950, tends to attribute this bias toward the center to a confluence of factors.[30] First, the British left the imprimatur of a strong central government, one that withstood the gradual devolution contained in the 1919 or 1935 Government of India Acts.[31] The legacies of colonial rule were not merely institutional but also ideological, constraining postindependence leaders with the straitjacket of "monolithic sovereignty" bequeathed by the British.[32] Second, the partition of British India and the fear of further disintegration provided ample justification for strengthening the powers of the central executive. The nature of India's party system in the early decades was a third centralizing force. The all-India presence of the Congress Party, which formed governments not only in the center but also the states, served to ensure that policy was coordinated rather than confrontational.

Much scholarship of political economy has likewise de-emphasized regional differences produced by federalism. Whether attuned to the country's "dominant proprietary classes,"[33] the oscillation between a "command" and a "demand" polity,[34] or the paradox of India's passive revolution[35] wrought by a political elite that sought development but wanted to avoid the "political costs of a direct attack on the existing social order,"[36] much of the work on political economy has looked at India as a whole. While many of these scholars allow for the importance of federalism, there is less explicit inquiry into how regional arenas contain neither the same configuration of regional actors nor the same kinds of state-society interactions. The possible consequence of variation in regional minicoalitions is a patchwork of political economies, each moving in its own direction and by its own logic. In much of this literature, there is an allusion to subnational politics, but the precise consequences of this fragmenting of state power remain only intriguing possibilities.

By contrast, political economy research trained at the state level, both comparative research and single case analyses, illuminates regionally specific patterns of development. This growing of body of work has helped to show how subnational differences in electoral politics, social structures, and individual state governments' relationships with the central government account for important regional variation in arenas as diverse as land reforms, industrial relations, economic liberalization, and the deployment of new technologies of governance.[37] For example, by looking at subnational variation in industrialization, Sinha challenges the conventional understanding that India is a failed developmental state. To explain uneven developmental outcomes, Sinha builds a story of "polycentric hierarchy," emphasizing the "interdependence of regional orders and national policy."[38] Variation in regional institutions and in the strategies of regional political actors then account for variable industrial investment across India. The insights of this subnational approach to studying Indian development and governance are an important counterpoint to scholarship that looks only at events in the national capital, and my own work builds on this research.

In concentrating more on provincial politics and less on center-province relations, I am influenced by Kohli's analysis of how subnational party organization and ideology have influenced social policies.[39] For Kohli, the locus of differential regional development regimes was the party system. But while parties and party ideology are undoubtedly important in explaining policy choice in some domains, particularly in social sector spending, there are also important exceptions. We might think of states governed by the same political parties that nevertheless pursued different social and economic policies, or the same policies but with different levels of intensity or effectiveness, as Desai demonstrates by comparing Communist-ruled West Bengal and Kerala.[40] Based on the case studies of this book, I argue that there are state-level development tendencies that endure despite changes in party structure, carried along by more persistent social foundations to political power. For example, in the state of Odisha, a rapidly changing party system from the late 1950s to the early 1990s masked more enduring continuities in the state's development focus. Throughout this long period, successive governments in Odisha have displayed a singular commitment to industrialization at the expense of agricultural transformation.

After several decades in which class had fallen out of scholarly favor, there has been a resurgent interest in analyzing class and social structure to illuminate Indian politics and political economy.[41] For example, we have evidence that regime type matters in determining the extent and nature of social policy and outcomes, particularly states' attempts at poverty reduction.[42] Those states in which lower classes and lower castes had more political influence pursued poverty reduction policies more consistently than those states that continued to be dominated exclusively by upper classes and upper castes. The more the lower classes and suppressed castes were politically mobilized and represented, the greater the likelihood that a state pursued redistributive policies. This is also the case with electrification. Where rural actors either acted on the state from the inside through rural political coalitions or from the outside through social movements, the state expanded rural electrification programs and lowered the cost of electricity for rural consumers. The absence of such pressures helps to account for the large swath of Indian countryside that remains unconnected to the grid.[43] Drawing from this literature, I take type of regime to refer to the class and caste composition of a state, expressed through formal political organizations like parties, but also evident in the strength of civil society actors and their relationship to the state.

In spite of calls to centralize electricity, which I discuss in Chapter 2, and despite the importance of electrification to New Delhi's agenda of planned industrial transformation, the Indian Constitution places electricity on the concurrent list, meaning that authority is shared between the center and the states. The central government provides the broad legislative framework within which subnational states function. As in the social spheres of education and health, and other physical sectors like irrigation and agriculture, in the electricity sector too it is the subnational government and not the central state that makes crucial decisions. Though all of the funds and some oversight rested with the central government through the allocation of revenues and the Planning Commission, the states nevertheless had considerable latitude to organize and execute these sectors to meet local and provincial agendas. This latitude was evident in how states allocated funds for rural electrification, in the tariffs they charged different classes of consumers, and in which consumer segments bore the brunt of electricity scarcities. One result of these differences was a persistent unevenness in infrastructural state building that helps to explain variation not just in development outcomes during the earlier period but also in economic policy making in the era of market reforms.

Electric Development and Distributional Conflict

Electricity has certain unique characteristics as a commodity: it must be consumed just as it is produced, the system as a whole has to constantly match available supply with demand, and capacity addition requires time. These qualities, combined with the fact that electricity is universally important, make this a particularly useful arena in which to investigate distributive politics. The allocation and pricing of electricity involve clear trade-offs between sectors and classes of consumers—true everywhere, but particularly where demand for energy habitually exceeds the capacity of the system. Scarcity forces the state, regulator, or system operator to make decisions that have profound implications for nearly everyone: industrialists for whom energy is a critical input; farmers who rely on electric-powered tube wells to pump water from subterranean reservoirs; India's famed high-tech ventures and call centers; and ordinary citizens, many of whom now live an energy-dependent life replete with computers, televisions, washing machines, and air conditioners. The outcomes of these decisions also have implications for the direction of economic development in India.

The literature on distributional conflict and development is extensive, among which the urban bias thesis represents one of the most influential variants of the last several decades.[44] The argument is that in the context of late development, states will extract surplus from the countryside to fuel development. By contrast, in industrialized states, the few remaining rural producers are the beneficiaries of extensive state subsidies. Since it was first formulated, others have challenged and added nuance to the original argument.[45] For example, Varshney contends that where democracy precedes industrialization, the tendency toward urban bias can be subverted.[46] The numerically superior rural sectors become empowered by democratic institutions to swing economic policy in their favor.

Just such a scenario unfolded in India. As the countryside's clout grew over the late 1970s and 1980s, agricultural producers secured a triad of important input subsidies (in addition to the subsidies for food prices, extension services, hybrid seeds, and credit). These subsidies—for irrigation, power, and fertilizers—were requisites of the Green Revolution. Of these, the subsidy for power had the clearest repercussion for nonagrarian sectors. Farmers began to use more power and they paid lower rates. In practice, many public utilities instituted a system of cross-subsidies to recoup their losses from providing cheap electricity to farmers, so industrial and commercial consumers were paying for the power that farmers used. Utilities dealt with insufficient supplies by curtailing access, called "load shedding"; in states where farm access was prioritized, other categories of consumers bore the brunt of this rationing. Whereas rural electrification programs were vibrant and aggressive in some states, implicating electricity deeper into the process of agricultural production, in other parts of the country, rural electrification was never so deeply subsidized and never became a developmental priority.

Based on an analysis of uneven patterns of electrification, I argue that the rural-urban dyad is underspecified, as others have also suggested.[47] Although India's farmers made great strides in redirecting development resources once they captured political power, the countryside's clout was very uneven across India. Examples of the beneficiaries include wheat and rice farmers in Punjab, Haryana, and Andhra Pradesh, and sugarcane farmers in western Maharashtra. However, large swaths of the countryside never reaped the benefits of the new deal for Indian agriculture.

Whereas the Indian central government used marketing boards and price controls to favor certain crops and groups of farmers, state governments used input subsidies and development expenditures to accomplish similar goals. In the case of electricity, the only beneficiaries were farmers in regions that were electrified and who could afford expensive equipment to sink tube wells for irrigation. So the new "rural bias" was limited both by class differentials within particular regions of the countryside and by differences in the level and nature of infrastructure development across subregions within individual states. The rural-urban dyad elides these marked, and perhaps no less politically significant, intrarural inequalities. As I propose in the following chapters, a more salient political cleavage may be the one that demarcates agrarian classes who move between the countryside and the state, directing public resources in their favor, and the larger majority of rural actors who have not profited from rural development programs. How electricity is allocated at the state level becomes a useful gauge of the development priorities of different state-level political regimes. The three case studies go further to ask why rural groups have been much more powerful in some states than in others.

Research Strategies: Qualitative, Comparative, and Historical

I should start by pointing out what this book does not seek to accomplish. This is not an analysis of the technical challenges in the Indian electricity sector. Rather I focus on the political factors that help to explain state-wise variation in rural electrification, and in turn how these patterns shed light on divergent experiences with utility privatization during the 1990s.

I take an approach that is qualitative, comparative, and historical. India's public utilities and government departments have produced a wealth of numerical data over the years, and I rely on much of it to buttress arguments throughout the book. However, many of the ways that variables are measured change over time, which makes quantitative analysis difficult. For example, once public utilities started to measure rural energy consumption based on the size of equipment rather than direct metering, the category of agricultural consumption grew disproportionately quickly and became a useful site to mask growing losses from technical failures and electricity theft. Data on village electrification are similarly problematic. For example, in a 1974 survey, thirteen of the seventy-seven villages that the public utility in Odisha counted as being "electrified," did not have a single consumer of electricity. The wires of the grid reached the village perimeter but no electricity flowed through them. Moreover, I wanted to understand whether and *how* politics impacted electrification; to get a textural sense of these processes, qualitative analysis of reports and documents was more useful than a purely quantitative assessment of the growth of electrification. There were additional problems of finding comparable data in each field site.

The book's approach is comparative in that it examines a single sector across three states within India. The subnational comparative method minimizes variation across units in the messy and complex terrains of society and politics.[48] Comparative analysis, in turn, helps to illuminate why economic structures and processes, access to public services, and lifestyles differ so widely across India. The advantages of comparing the three states of Maharashtra, Odisha, and Andhra Pradesh are threefold. First, the three together represent a variety of experiences with privatization during the 1990s. Whereas privatization of the distribution utility was carried out first in Odisha, the policy was not considered seriously in Maharashtra, and in Andhra Pradesh, a government attempt at privatization was ultimately foreclosed. Second, the project has geographic breadth by engaging with states from the western, southern, and eastern regions of the country. And third, in the selection of Odisha, I hope to expand the conversation about politics and development in India by including a part of the country that has been overlooked in most comparative studies.

In making sense of variation in how these three states crafted policy during the first decade of market reforms, I was compelled to work backward in order to understand disparities in the process of infrastructural expansion during the prior three decades. The historical aspect of the book attends to what I believe were foundational decades: the 1960s to the 1980s. These decades encompass the period in which newly wrought linguistic states created the institutions and cemented some of the political patterns that would mark these regional political economies into the present.[49] As historian Ramachandra Guha has noted, many of the important political, social, and economic histories of postcolonial India are yet to be written, overlooked in the gaps that form between academic disciplines. While historians of India have focused on the well-documented colonial past, sociologists and political scientists have a bias for the present that generally confines their analyses to smaller temporal frames.[50] Linking the contemporary process of subnational market reforms with historical patterns of infrastructural development is an effort to bridge such gaps.

Focusing on state-level political economy has also highlighted another feature of postcolonial India, namely the extremely uneven nature of the state as a record keeper. As the first research task for this project, I set out to understand the history of electrification in India by looking through the archives of the SEBs, which were the largest public utilities in the sector from the 1950s to the early 2000s. Each SEB was obligated to produce an annual report that documented its financial, technical, and operational activities for the preceding year, deposited in their respective state legislatures and available for legislators to query. If I could locate these, I believed I could piece together how states had chosen to finance and expand their electric grids from the time they were created. In each of the three states that I focus on in this book, I was met with a distinct experience. The search for these records supplied my first lesson in the fragmented nature of the Indian state, although I did not appreciate the value of the instruction at the time.

In Mumbai, the capital city of Maharashtra, I was welcomed to the SEB offices by the chief secretary of the Maharashtra State Electricity Board (MSEB), who escorted me through well-appointed offices located in the heart of the city. In a large room, I found bound volumes of the annual reports of the MSEB, organized chronologically and with no missing years. I was given the use of a vacant desk and invited to peruse the volumes that day and for as many days thereafter as I needed. I spent the bulk of my time during that first trip reading through the annual reports, taking notes, and meeting with bureaucrats in the financial and administrative divisions of the SEB. I subsequently found nearly complete sets of annual reports from MSEB and other energy sector documents in the library of the Gokhale Institute of Politics and Economics in Pune.

In Andhra Pradesh's capital city, Hyderabad, in the former headquarters of the Andhra Pradesh State Electricity Board (APSEB), now occupied by the successor entity, the Andhra Pradesh Power Generation Corporation (APGENCO), I was again warmly received by a senior bureaucrat, who assigned a clerk to take me to the institution's library on the ground floor of the building. A generous librarian introduced me to the several aisles of floor-to-ceiling bookshelves on which she guessed I would find various issues of the annual reports. I spent the first two days of my time in the library assembling an archive out of scattered, dusty materials that seemed to be stored almost incidentally. In the end, I was able to find copies of most but not all years from the time the utility was established in 1959, following the formation of the new linguistic state of Andhra Pradesh, to the early 1990s, which had been my goal. The librarian very kindly assisted me to photocopy certain pages from these records by assigning the library's peon to the task. I was uncertain that every page I wanted would indeed be photocopied, but decided that if anything ended up missing, I could return to the public record that I had just finished assembling.

In Bhubaneswar, the capital city of Odisha, my encounter was again different. My experiences in Maharashtra and Andhra Pradesh had convinced me that the most likely repository of a complete record of annual reports would be the offices of the SEB or its successor entities. In Odisha, which had been the first of India's states to dismantle its vertically integrated utility into separate functional units, the headquarters of the former Odisha State Electricity Board (OSEB) were occupied by the Grid Corporation of Odisha (GRIDCO), which continued under state ownership even after other pieces of the utility were privatized. There, I met with the chief managing director, the highest officer of the utility, who welcomed me to use the utility's library but expressed skepticism that I would find what I sought. I assured him that the other SEBs I had visited had kept these records, and the OSEB would likely be no different. His skepticism was warranted. Housed in a small bungalow behind the main campus of the utility, the library consisted of several rooms that included a modest room whose center was almost entirely taken up with a large conference table and against whose walls I found roughly six bookshelves. Like the APSEB library, this one too was meant for the use of utility employees, and the main collection of materials consisted of engineering textbooks. From a small room off of this conference room, the librarian emerged with a stack of roughly seven softcover reports that represented all of the annual reports she could locate. These were a smattering of annual reports mostly from the 1980s and 1990s, with nothing from the 1950s, 1960s, or 1970s.

On a subsequent research trip to Bhubaneswar, I again sought out the annual reports of the erstwhile OSEB. I looked in several places, including the state archives, the largest state-funded library in the city, the Directorate of Economics and Statistics, the library at the Nabakrushna Choudhury Centre for Development Studies, and finally the library attached to the Odisha Legislative Assembly, entry to which I secured thanks to extremely helpful colleagues in Bhubaneswar. My search through all of these libraries, record rooms, and archives proved futile.

If these differences tell us something about the record-keeping role of the state, then they also tell us something about the varying administrative capacities of India's subnational units. The "regional imbalances" of which New Delhi's planners and politicians were so cognizant at independence, and which they wanted to efface over time, were reflected not merely in the wildly different per capita incomes and levels of industrialization found across the former British provinces and princely states but also in the political and administrative machinery of the newly created subnational states. Like other measures of inequality, this difference in administrative capacity seems to have persisted over time rather than narrowing.

In addition to the utility archives that I was able to find in each state capital and the central government documents in New Delhi, I also used a range of other state-level planning and public-sector documents to reconstruct the pace and pattern of electrification in each state from the 1960s onward. Interviews with bureaucrats, activists, and politicians connected to the power sector provided valuable contextual details that were missing from the written record.

Outline of the Book

In Chapter 2, I examine the institutions and ideologies that emerged from New Delhi. During the 1930s and 1940s, figures like Nehru and Ambedkar embodied the view that science and technology could work through the levers of industrialization to resolve India's problems of poverty, underdevelopment, and inequality. Emboldened and inspired by Lenin's Russia, they and others in the worlds of planning and engineering believed that state-financed electrification would be the means of transforming the productive capacities of the country, as well as ensuring India's arrival as a fully modern nation. Running through these early debates were tensions around how electricity should be organized. Although there were prominent supporters of a centralized system, the electric grid was organized along federal boundaries. The late 1950s through the 1980s represent the heyday of provincial public utilities. Partly in response to acute agrarian crises that revealed India's continued dependence on foreign largesse, particularly in food aid, energy again became critical to a new vision for economic transformation. Accessing India's ample groundwater resources through electrified tube wells was a cornerstone of India's Green Revolution strategy, for which the government in New Delhi developed new incentives to support rural electrification programs in the provinces. Throughout this period, the central government gradually wrested authority for energy and electricity from the states, first to stimulate more rural electrification and later to expand generation supplies and interconnect the various state and regional grids. Finally, in the third period starting in the early 1990s, in response to the continued scarcity of electricity supplies and as a part of a larger turn toward the market, New Delhi created new rules to facilitate the entry of private capital into a sector that hitherto had been the purview of the state.

Chapter 3 analyzes the politics of electricity and electrification in Maharashtra. Of the three cases, Maharashtra provides the strongest example of how a politicized peasantry wielded the instruments of democratic institutions and an expanding state sector to channel resources to the countryside, although farmers in just a handful of districts cornered most of the gains. Even before independence, agrarian elites organized in the Maratha caste cluster began to command a large measure of influence in the nationalist movement in western India. After independence, it was from the districts just east of Mumbai that members of a new political class emerged from the rich farmlands that lent themselves to sugarcane production. Extensive rural electrification programs heightened the productivity of these districts, and the state's subsidized electricity along with crucial support for cooperative farming and credit institutions bound the political order in Maharashtra to an emerging class of commercial farmers and agro-industrialists, known as the "sugar barons." In the period of market reforms, the regime of rural subsidies and the ability of Maharashtra's large farmers to protect that regime made utility privatization an unattractive solution to the problems in the sector, although some in the policy community believed Maharashtra would be an ideal candidate for privatization. In its place, the state government adopted policies that were nominally market friendly but continued to protect the interests of powerful rural actors, for example, through the promotion of a bagasse industry.

Chapter 4 turns to the eastern state of Odisha. I argue that the extremely low rates of rural household electrification and rural electricity consumption are emblematic of a general dearth of development resources and attention to agriculture. Despite having a large rural population and large lower caste and tribal communities, Odisha continues to be governed by a political establishment with more extensive ties to urban and industrial pursuits than to the countryside. In the period of market reforms, Odisha emerged as the first state in India to pursue the kind of utility privatization and restructuring that already had been implemented in most Latin American and many African and Asian countries. Because the state in Odisha had extended fewer development resources to rural populations, the contraction of its public sector was more easily accomplished. Privatization also served as a signal to private investors that the state would provide a hospitable investment climate, and thus advanced the strategy of extractive industrialization that Odisha's political leaders had long favored.

Chapter 5 focuses on Andhra Pradesh. By the 1990s, although the state's chief minister, N. Chandrababu Naidu, was widely hailed as India's most private-investor-friendly and reformist political chief, an active farm sector ultimately forced the state to adopt backdoor market reforms. Earlier, the provision of electricity to the rural sectors made the greatest gains when the political landscape in the state became more competitive starting in the late 1970s. Throughout the 1980s, populist promises made in the context of newly competitive party politics continued to spur electrification and reduce electricity tariffs for farmers. In the 1990s, Naidu wagered his political capital on utility privatization only to have it opposed by the state's influential farming communities who benefited from subsidized electricity.

The conclusion extends the arguments of the previous chapters in two directions. The first and larger of the two tasks considers how well the arguments of the book travel to other states in India. To that end, I consider the politics of electrification in Punjab, Tamil Nadu, and West Bengal. The last, short section of the book reflects on events in the past decade, after passage of the Electricity Act in 2003.

2

Electricity as New India's "Strategic Railway"

Building the Infrastructural State

The political and economic history of the railway system in colonial India presages both metaphorically and practically the creation of the electrical grid in India after independence—with one important difference. While the railroads become the purview of the central government, oversight of electricity resided with India's state governments. In British India, the construction of a vast rail network was powerful, both for the material changes it wrought and its symbolic valence. As a recent reexamination of the production of the colonial economy notes, of all the public works that were critical to the colonial enterprise in India in the nineteenth century, the railways came closest to "distilling and communicating the ideology of colonial state space."[1] So saturated with meaning, the railways invited multiple and contradictory interpretations of their effects. For the British administrators, the railways represented the metropole's colossal efforts to improve the colony. Proponents also underscored their promise to create a more homogenous space that would allow opportunities for trade that earlier had been confined to the coasts to penetrate the Indian interior. But for Dadabhai Naoroji and other early economic nationalists, the railways represented the exploitive and uneven effects of the colonial enterprise. Tracks were laid to serve colonial—not indigenous—interests, and the extent of the rails outlined the scope of British power, incorporating those areas where troops could most quickly be delivered and from which natural and other resources could more easily flow outward. These critics emphasized how the railways had created a new overlay of uneven development and new avenues for Indian indebtedness to England.[2] While uniting some parts of the subcontinent with a modern transportation infrastructure, the railways simultaneously divided India into regions that were accessed by modernity and other regions isolated from the reach of the colonial and the modern.

During the early decades of railroad construction in India, from the 1840s to the late 1860s, private British companies built railroad lines with capital raised in London. Throughout this period, the Indian railroads constituted the largest single investment outflow of British capital, eclipsing the sums going either to other parts of the empire or to other economic sectors.[3] The East India Company and, after 1858, the government of India backed these speculative investments by ensuring that private companies would receive a guaranteed rate of return. For economic nationalists, this guaranteed return was also the subject of criticism, held up as emblematic of how the colonial state directed profits from successful investments to private investors in the metropole, but socialized the costs of failed investments by pledging state coffers raised through local taxes.[4] By the late 1860s, after several decades of critique from nationalists and intermittent famines that were analyzed as the result of deficient transportation to famine-prone districts, the government of India assumed responsibility for the railroads, taking up the leases of private companies as they lapsed and undertaking the construction of new rail lines. During this period of direct government control of the railroads, the colonial administrators understood well their importance for the broader British mission in India, a mission that would serve economic advancement as much as it would serve political control of the vast regions of the subcontinent under their aegis.[5]

In sheer physical size, capital investment, and symbolic authority, the project of electrification was postcolonial India's analogue to the railroads. Like railroads, electrification too would be a state-directed project with far-reaching material consequences that was also critical as an emblem of a new economic future. The ambit of the project was correspondingly large, from the upstream generation of electricity in massive dams, nuclear generators, and coal-fired thermal plants to the downstream distribution of electricity to fuel India's new industries, irrigate farmlands, and illuminate public spaces and private homes. Like railroads, early utility companies in British India were largely privately owned, but the government fixed the rate of profit that these companies could earn. Also like the railroads, electricity was a new investment node with a sizable presence on the emerging Bombay stock exchange prior to independence. After independence, during most five-year and annual plan periods, the spending on electricity was generally larger than for any other category of spending, echoing the vast resources sequestered for railway construction in the colonial period. And, finally, like the railroads, the project of electrification was imagined to be a universalizing agent, a means to unite and develop the resources of diverse and disparate units of India; however, it produced a reality that belied its promise. The nationalist faith that electricity and electrification could be tools to promote more balanced and even development has so far remained unrealized.

After the earlier privatized systems gave way to state ownership, both the railroads in British India and postcolonial electrification became central to the project of infrastructural state building. The railways and the power grid are not merely infrastructure in a literal sense but also in the way that Michael Mann uses the term in his theory of state power.[6] Aside from extending the reach of the state outward, as Mann suggests, these technologies of movement and communications also served an important function in strengthening the symbolic authority of the new state to govern. Especially for India's resource-constrained postcolonial democratic state, monumental public works were critical pieces of the project, not just of state building but also of nation building.[7] As the key monumental structure, the big dam showed the ability of the modern state to "rearrange the natural and social environment" and thereby demonstrate the strength of its technoeconomic might.[8] The impact of dams was clear, especially for anyone who could witness firsthand how they transformed the local environment. But the electricity that dams were built to produce spread along wires to affect individuals at the level of the everyday. Electricity became a conduit for the nationalist project in India, which allowed the "sights and sounds of the nation," to invade public and private spaces alike.[9]

But by ceding the management and expansion of electricity to subnational state authorities, the Indian government effectively ensured that this important piece of state building would be a heterogeneous process rather than a homogenizing one. Some states put an early emphasis on rural electrification as a way to increase productivity, but through this conduit, rural households also gained broader access to electricity. Analyzing electrification in India throws into relief the regional and state-level political economies of development that effected different patterns of investment and infrastructural state building.

As the following chapters focus on the politics of electrification at the provincial level—in the states of Maharashtra, Odisha, and Andhra

Pradesh—this chapter delineates the broad, national-level institutional context in which the provincial politics of electrification took place from just before independence until 2003. I engage the debates about the institutional blueprint for the electricity sector over a long time period, beginning with the years straddling independence when a new legislative architecture for the electricity sector established public utilities in each state in India and the Central Electricity Authority (CEA) in New Delhi. Subsequently, from the 1950s to the 1980s, the State Electricity Boards (SEBs) emerged as the preeminent institutions in the sector. Over this time, state governments began to use electrification and SEBs as tools to shape particular development strategies. In some states, rural electrification provided a fillip for agrarian productivity, contributing to India's Green Revolution, while in other states, electricity remained confined to urban and industrial consumers. During this period, the central government also expanded its reach into the sector, but took a cautious rather than radical approach in doing so. In the final period, from the early 1990s to the passage of the Electricity Act of 2003, the overall center of gravity of the electricity sector started to shift toward private capital and market logics. The 2003 legislation, the genesis and form of which I discuss later in this chapter, marks the beginning of a new phase of India's energy and electricity history, one whose shape is still inchoate.

Throughout the sixty-year period from the 1940s to the early 2000s, two issues were debated: (1) whether the sector should be owned by the state or private firms, and (2) which level of the state—the center or the provinces—should exercise the greater share of control. With respect to the first issue, India has come full circle over sixty years. At independence, India's electricity sector was populated by hundreds of privately owned utilities. Over time, the majority of these—although with important exceptions—were nationalized and integrated to form a series of state-owned grids, organized according to India's federal structure. By the 1990s, the central government re-formed the laws governing electricity to create more opportunities for private firms, returning to a scenario that recalled the end of the colonial period.

On the second debate about state versus central control, the change has been in a single direction, toward increasing—but never complete—central control. Institutional patterns inherited from the colonial period left electricity as one of the preeminent domains of the subnational state. Over time, though, the central government emerged as a major actor in the electricity sector, first as a financier and later as a generating company, a transmission company, and a distribution company. From a set of provincial-level, publicly owned electricity networks, India has moved closer to having a nationally integrated grid structure crowded with private and public firms alike.

Electricity in the Nationalist Imagination

In the decades just before independence, the role of science and technology was at the core of debates about India's future; within this broad set of debates, electricity was discussed as the backbone of plans for growth and industrialization. On the one hand were those who held a fervent belief that science and technology would transform existing conditions. Such views were particularly common among the Western-educated nationalist elite in the last decades of the colonial period.

The counterposition, on the other hand, emphasized the dark consequences that science and technology would serve, predicting that modern progress would displace labor, degrade the environment, and threaten social stability—views held most famously by M. K. Gandhi. For Gandhi, the Western route to material progress was at odds with the path to moral progress. In a 1916 lecture on development, he remarked that "the smoke and the din of mill chimneys and factories . . . are held to be symbolical of material progress," and yet he was sure that "they add not an atom to our happiness."[10] Industrialization, according to Gandhi, negatively affected not just the physical environment but also the mental worlds of the people it touched, as well as the social relations among them.

Gandhi's antipathy toward electricity was slightly distinct from his opposition to industrialization in general in that he clearly perceived the connection between state power and electrical power. He predicted that electrification would bring an end to local autonomy.[11] From a 1934 article in the journal *Harijan*:

> While it is true that you will be producing things in innumerable areas, the power will come from one selected centre. That, in the end, I think would be found disastrous. It would place such limitless power in one human agency that I dread to think of it. The consequences, for instance, of such a control of power would be that I would be dependent on that power for light, water, even air and so on. That, I think, would be terrible.[12]

Gandhi foresaw, then, precisely the uses that the electric grid could serve in the consolidation of the Indian state's infrastructural power. But while this was a dreadful possibility for Gandhi, for Nehru and other centrally minded planners, it was a longed-for future. Gandhi conceded that electricity transmitted through a grid could make decentralized industrialization more economically viable, but his overriding worry was that such a route to development might threaten the autonomy of local communities. As respected a statesman and political organizer as Gandhi was, his thoughts on India's economic future were marginalized even at the height of his political and social influence. For many, Gandhi's "modernity" looked only backward to an imagined era of the self-sufficient village powered by human and animal labor, and not far enough forward to a new world order built on electricity and harnessing a nation's natural, human, and technological resources on the world stage.[13]

As a self-proclaimed disciple of Gandhi, Jawaharlal Nehru's views on industrialization, modernization, and the roles of science and technology in these processes nevertheless were almost perfectly opposite to Gandhi's. Among those who emphasized science's redemptive promise, Jawaharlal Nehru stands out both because of the intensity of his convictions and because, as India's first prime minister, his convictions were so consequential for postcolonial India. His views of science and technology in general are well-known,[14] but we get the clearest sense of his perspective on electrification in particular from how he analyzed the Soviet experience. Nehru's *Glimpses of World History* comprises a series of letters that he wrote to his then teenage daughter, Indira, during one of his many imprisonments for anticolonial activities. As its title suggests, the book frames a long world history, with each chapter (originally written as a letter) devoted to some episode from history, from early considerations of the Greek city-states and Alexander's imperial ambitions to later chapters on international monetary systems and the rise of fascism in midcentury Europe. The chapter on the period of the Bolshevik Revolution, which Nehru penned in July 1933, just fifteen years after the event, is an enthusiastic review of the structural transformations that the new Russian polity had accomplished with breathtaking speed. Nehru invokes Lenin's famous phrase to the Eighth Soviet, that "electricity plus soviets equals socialism," as the heart of this triumph.[15] For Nehru, part of the genius of the Russian electrification program was its ability to resolve contradictions between the urban core and the Russian hinterlands, and between industry and agriculture. Another part was in its noneconomic side effects. As Nehru admiringly notes, "the peasants, whose villages were lighted up by electricity and much of whose farm work was done by electric power, began to get out of the old ruts and superstitions and to think on new lines."[16] Electrification shined the "lamp of Ilich" on what were considered backward peasant ways of life.[17]

Russian electrification also signaled the ascendance of a new political class—the engineers—and a new commitment to technology as a panacea for a range of problems confronting Russia after the war; it promised a "technological utopia."[18] The engineers of GEOLRO (State Commission for the Electrification of Russia), allied to the highest echelons of the Communist Party and with the explicit patronage of Lenin, expanded the ambit of their mandate to offer a holistic industrial plan for Soviet Russia.[19] Electrification, then, was a prototype for centralized planning in general.

Nehru was not alone in believing that electricity would transform India. A similar view was held by M. Visvesvaraya, an engineer by training who spent the first fourteen years of his career in Bombay Presidency, designing and improving municipal water supplies and canal waters. After leaving the British service, Visvesvaraya spent five months traveling throughout Europe and North America. He later credits these travels with convincing him that the task of governments must be to lay the foundations for industrial development.[20] Returning to India, he worked briefly in the Nizam's territories on a flood-control plan for Hyderabad, which had been inundated by rising river waters in 1908, before joining the administration of princely Mysore. There, Visvesvaraya was actively engaged in some of the earliest experiments with state-directed industrialization, irrigation development, and electricity production in India, first as chief engineer and later as *dewan*, the highest administrative position in the state.[21] Under Visvesvaraya as chief engineer starting in 1909, electricity production in Mysore expanded from 13,000 hp in 1909 to 25,000 hp by 1919, with a corollary increase in the government's electricity tax collection from Rs. 1.7 million in 1911–12 to Rs. 2.4 million in 1918–19.[22] By 1948–49, Mysore's revenue from electricity had climbed to Rs. 13.3 million. By expanding electricity, Mysore was able to rapidly expand its industrial base, and by the time Visvesvaraya stepped down as *dewan* in 1918–19, the state had several new state and private companies, including a soap manufacturing plant, chrome tanning and metal factories, and new printing presses, all benefiting from Mysore's public electric utility.[23]

Visvesvaraya penned a very early economic plan for India in 1936, which consists of two broad parts: in the first, he assesses India's development, and in the second, he lays out a ten-year program to rapidly increase standards of living.[24] In a chapter devoted to India's power supplies and transportation infrastructure, Visvesvaraya reproduces a table from a 1933 issue of the *American Economic Review* that measures productivity by the daily output of work in India, the United States, Japan, and other industrialized countries. In addition to having far lower output per capita than all the other countries on the list, India's work output is dominated by human labor, whereas in the

United States, the dominant labor output comes from coal usage. In concluding his analysis in this section, Visvesvaraya suggests that India's leaders must "utilize to the best advantage the technical results of modern science and thus pave the way to the removal of the reproach conveyed by the low position occupied by India in this table."[25] The relative poverty of India's energy production and consumption were, for Visvesvaraya, not only a stark measure of just how far India must travel to catch up to the industrialized world but also, as the earlier quote suggests, a cause for moral reproof.

Indian planners were not alone in expressing anxiety about their country's electrical backwardness. Just one decade before Visvesvaraya drafted his economic plan for India, the British Parliament passed the Electricity Supply Act of 1926, which created for the first time a single national electric grid in Britain. The legislative drafting committee marshaled support for the bill by arguing that the decline of British industrial power was fundamentally a failure of electric power. Supporters of the bill assembled a ranking of cities and regions that included California, Chicago, and Great Britain according to per capita electricity consumption, in which Great Britain appeared last. Such evidence stoked fears that Britain was losing its economic edge, and allowed the bill's proponents to garner broader political and public support for change.[26]

Both Nehru and Visvesvaraya considered rapidly expanding production and broad access to electricity vital for overall industrial transformation. As chief engineer and later *dewan*, Visvesvaraya had been responsible for substantially increasing electricity production in Mysore through public investment. But in his plan for India as a whole, Visvesvaraya did not consider the modalities of electrification, whether it should be a state or private-sector project, and if the former, whether it should be a central or provincial task.

More details were provided in the last decade of colonial rule by two parallel processes of economic planning, one carried out by the government of British India and the other by the Congress Party's National Planning Committee.[27] These two efforts serve as another reminder that far from originating with the postcolonial state, both the ideas and the early blueprints for economic planning in India originated in the colonial period.

The colonial government wanted to outline what was intended to become a full-fledged planning apparatus at the war's end. In this, the colonial government was inspired both by the reconstruction planning in Europe (begun in England in 1941) and by the global movement toward an increasing role for the state in the economy. The government instituted a series of committees organized by economic domain that were meant to recommend future policies. B. R. Ambedkar was assigned as chairman of the committee on public utilities and electricity. By this point, Ambedkar already had held various positions in the government, including that of labor minister, and he would later be appointed chairman of the Drafting Committee of the Constituent Assembly. Ambedkar is best remembered as the most important leader of India's *dalit*, or former "untouchable," communities, as well as for his contributions to the democratic constitution and caste equality. But for Ambedkar, mapping out the political and technical details of India's industrialized future were not disconnected from "the caste question."[28] Rather, in his analysis, urbanization and industrialization provided the best hope to overcome caste hierarchies, social discrimination, and inequality. Like Nehru and Visvesvaraya, Ambedkar was certain that electricity would be critical to India's future. As chairman of the committee on public utilities and electricity, Ambedkar made a series of policy speeches in the central legislative assembly in which he elaborated a perceived set of relationships between electricity, industrialization, and poverty:

> It is necessary that those who are placed in charge of the subject should have the fullest realization of [electrification's] significance and its objective. If you agree with me in this, I will request you to ask yourselves the question, "Why do we want cheap and abundant electricity in India?" The answer is that without cheap and abundant electricity no effort for the industrialization of India can succeed. This answer brings out only a part of the significance of the work this Committee has to undertake. Ask another question, "Why is industrialization necessary?" and you will have the full significance made clear to you at once; for the answer to the question is, we want industrialization in India as the surest means to rescue the people from the eternal cycle of poverty in which they are caught.[29]

After laying out the significance of the committee's work, Ambedkar further identified the two most critical and potentially divisive questions that needed to be resolved as the committee framed its policy for the sector: state versus private ownership and provincial versus central control. The comparison between the emerging electrical grid and the colonial railway system was very apparent to Ambedkar. In adjudicating the first issue, Ambedkar suggested that the committee should think of electricity as a "strategic Railway," and as "an undertaking which must be started without consideration of immediate profit."[30] By arguing that electricity must be developed quickly and without thought for profit, Ambedkar was articulating a rationale for state ownership. Only state entities—or private companies with very large state subventions—would be willing to forgo immediate profits to fulfill longer-term needs. By 1945, Ambedkar was unequivocal in advocating state ownership, noting that "there are very few opponents of State ownership and State control who do not make an exception in the case of electricity."[31] His language indicates that he faced little opposition to this view.

More than whether or not the state should own the grid, Ambedkar seemed concerned about which level of the state—the central or provincial governments—should do so. In the allocation of legislative subjects between the central, provincial, and concurrent lists, the Government of India Act 1935 had placed electricity in the third category. Prior to independence, the role of the central authorities was limited to laying out the broad rules for the sector while the provincial governments collected electricity taxes, granted licenses to private utilities, and in some cases, constructed government-owned generation and distribution grids. By the late 1930s, however, the war effort made it imperative for the central government to have more control over this strategic resource.[32] In 1938, the central government began compiling electricity statistics, first published in the December 1938 issue of the *Monthly Survey of Business Conditions in India*, a bulletin released by the economic adviser to the government through its statistical research branch. The government's press notification justified this new collection of data as being in keeping with emerging practices in "many countries as they are considered a good index of general economic conditions."[33] Some versions of these data were being assembled already, but only on an annual basis and in a disaggregated manner through the various provincial governments. In 1938, realizing the increasing importance of electricity for governance, the central government's statistical branch began to bypass the provincial governments and approach electric companies directly.

The centralization and standardization of data collection continued in 1941 with the installation of an entirely new bureaucratic post, the electrical commissioner. The post was recommended by the Ministry of Supply Mission, which reported that "an efficient and adequate supply of electrical energy is essential to War production."[34] One of the electricity commissioner's tasks was to coordinate the disparate private electricity generation and distribution utilities to safeguard electricity supplies to ordnance factories and other industries connected to waging war. Underscoring the relationship between electricity and war, the office of the first commissioner, H. M. Mathews, was located in Calcutta alongside the director general of munitions production, and was granted statutory powers under the Defense of India Rules. The second, and in some ways more important, task of the commissioner's office was to assemble data on electric utilities across India from private, provincial, and municipal licenses in both the British provinces and the native states.

So into the 1940s, the central government's role was restricted to data collection with some limited coordination functions. Ambedkar, however, advocated greater centralization. In speeches to the policy reconstruction committee in 1943 and 1945, Ambedkar worried that the lack of central control would hinder the overall development of electricity. During the two years that lapsed between those two meetings, Ambedkar also became closely involved in the ongoing plans for a multipurpose river project in the Damodar River valley in central and eastern India. Modeled on the Tennessee Valley Authority (TVA), the Damodar Valley Authority (DVA) was being planned as a multipurpose vehicle to control annual flooding, increase irrigation, and generate hydroelectricity. For Ambedkar, the appeal of the project was in how it crossed provincial boundaries. Describing the project in 1945, Ambedkar again invoked the colonial railway system when he remarked:

> We have not taken sufficient account of the fact that there is no difference between railways and waterways, and if railways cannot be subjected to provincial boundaries, neither can waterways, at any rate those that flow from province to province. On the contrary we have allowed our constitution to make a distinction between railways and waterways, with the result that the railways are treated as Central, but waterways are treated as Provincial. The disadvantages of this error are many and obvious. To give one illustration, a province needs electricity and wishes to utilize its water resources for the purpose but it cannot do so because the point at which water can be damned lie [*sic*] in another province which being agricultural does not need electricity and has no interest in it, or money to finance the project, and would not allow the needy province to use the site. Complain as much as we like, a Province can take such an unfriendly attitude and justify it in the name of Provincial Autonomy.[35]

Ambedkar's dim view of the Indian village is well known. As he famously put it, the village was nothing but a "sink of localism, a den of ignorance, narrow mindedness and communalism."[36] From the previous quote and from other remarks he made in the course of reconstruction planning, what becomes clear is that for Ambedkar, the province was equally problematic as an obstacle to progress, particularly for infrastructural development. His vision for development was calibrated to a national scale to ensure that vital resources were shared across the national space rather than hoarded in particular locales. Since electricity required either river waters for hydroelectric production or coal deposits for thermal production, Ambedkar believed that without central government intervention, electricity developed at the state level would be necessarily uneven, constrained by the way that these natural resources were distributed across provinces. In a committee meeting held in April 1945 to discuss the prospects for a multipurpose project on the river Sone, a large tributary of the Ganges that cuts through present-day Bihar, Uttar Pradesh, and Madhya Pradesh, Ambedkar again argued that infrastructural development projects must break out of their local and provincial box.[37]

Others in the administration felt similarly. B. G. Ghate, the undersecretary of planning and development under the last colonial government, also felt that irrigation and electricity would be hindered by provincial control. For him, the TVA represented an ideal. Recent scholarship has pointed out that the TVA inspired river valley projects in many parts of the world, also spreading the high-modernist belief in the limitless capacity of science and technology to tame nature.[38] But for Ghate, the brilliance of the TVA lay not only in how it controlled the physical world. Rather, the TVA allowed the central government to harness and control local politics, and bend these to serve a development vision imagined in Washington, DC, rather than in the provincial capitals of Nashville, Tennessee, and Jackson, Mississippi. On All-India Radio in April 1945, Ghate narrated "The Story of the T.V.A." to praise these multiple achievements. As he extolled, "running as it did through seven states . . . no individual state could undertake any big project on account of the financial, constitutional, technical and administrative difficulties involved."[39] In creating the TVA, Ghate argued that the US Congress effectively overcame a collective action problem that is endemic to natural resources shared across multiple political jurisdictions.

One of the intellectual forces behind the DVA was another man of science, Meghnad Saha, who had also served on the Congress's National Planning Committee as chairman of the subcommittee on power, fuels, and river training. On that committee, and through speeches and lectures in science and technology forums in Calcutta and New Delhi, Saha championed the idea that India's progress required the careful bending of nature to suit modern needs, in particular to produce electricity that could power industrialization. In a speech to the National Institute of Sciences, Saha virtually equated India's poverty to the underdeveloped state of its electricity resources, for which he blamed the greed of British companies and the incompetence of British administrators.[40] Like Ghate and Ambedkar, Saha too anticipated the problems that would come from the fact that provincial governments had so much control over a resource he believed should be politically centralized in New Delhi. He blamed the 1935 Government of India

Act for pandering to narrow interests by bifurcating authority between the central and provincial governments, thereby erecting considerable obstacles to the national planning effort.[41] The National Planning Committee's Subcommittee on Power and Fuel imagined a scenario in which the electricity sectors of Indian provinces were harmonized by empowered central agencies that would craft policy for the country as a whole rather than individual units or sectors alone.[42]

By the late 1940s, many planners and bureaucrats had come to suspect a contradiction between nation-level plans and the realities of provincial autonomy granted by the 1935 Government of India Act. Ambedkar's committee, for example, had to operate within the legislative ambit provided by the 1935 act, which restricted the central government's role to broad rule making in the electricity sector. This also was true for planners involved with the DVA, as well as in many other proposed arenas. In a historical account of flood control policies in the Mahanadi River valley in eastern India, D'Souza uncovers internal communications in central agencies that show British administrators were keen to increase the central government's authority while also remaining sensitive to the interests of provincial governments, which were now in the hands of elected legislatures. As one British bureaucrat in the Finance Ministry lamented in 1944, "almost the entire field of development lies in the Provincial sphere, e.g. education, agriculture, canals, drainage, embankments, water-power, communications . . . But postwar development on so ambitious a scale as is now contemplated involves strong Central direction, control and coordination in a field which is almost entirely provincial."[43] Many were aware of the political barrier to national development but were seemingly unable to overcome it. These limitations continued into the postindependence period, as the prevailing balance of power in favor of the states became cemented in the Electricity (Supply) Act of 1948.

India's Electricity System: Publicly Owned and Provincially Controlled

In 1948, India's leaders inherited a country with a miniscule amount of total electrical power (1,713 MW, or less than 2 percent of India's current capacity) and with only about half a percent of its villages electrified. Although the earliest drafts of the 1948 legislation were prepared before independence, the seriousness with which the subject was debated, with many detailed discussions of the financial, technical, and administrative aspects of electricity over several days, provides a record of what the makers of India's Constitution believed to be the most important and controversial aspects of electrification. In presenting their views and justifying their positions, the debaters explicitly drew on international experience and prevailing wisdom. They were ecumenical as well, referring to the centralized investment allocation and five-year plans of the Soviet Union, the United Kingdom's nationalized electricity system, and the massive public works of the United States. Just as it had for Nehru, the Soviet example cast a powerful shadow on the Indian constituent assembly. Lenin's elegant arithmetic of modernization—that communism was the sum of soviets plus electrification of the whole country—was invoked like a mantra throughout the debate in 1948.

During the colonial period, the Indian Electricity Act, 1910 had laid out the rules by which private firms were to be granted licenses to supply power. India's electricity sector was composed of hundreds of private supply and distribution companies located almost exclusively in cities and larger towns and the industrial regions surrounding them. While the majority of these were British owned, there were a few prominent Indian players, notably the Tata conglomerate. We get a sense of India's energy resources just before independence from data assembled by the Chief Electricity Commissioner's Office in 1944 (see Table 2.1).

Utilities operated in all of the provinces and also the largest of the native states, but the amount of capacity varied substantially. Among the British provinces, 58 percent of the total capacity of the system was located in just Bombay and Bengal. Among the princely states, 42 percent of the total capacity was located in Mysore. The fuel mix was also substantially different across regions, from the exclusively hydroelectric system of Mysore and regions with substantial hydroelectricity—like Bombay, Punjab, Madras, and Travancore—to Bengal, which had an almost entirely thermal system. Ownership structures also differed considerably. In the south in particular, in both princely states (Mysore and Travancore) and British provinces (Madras), governments had invested public resources to expand hydroelectricity. By contrast, the two largest electricity systems of Bengal and Bombay were almost exclusively in the private sector.

As the cities and larger towns were becoming increasingly well lit from the 1910s to the 1940s, smaller towns and villages in between were largely untouched by this new technology. Constituent Assembly members criticized the privatized system as being equal neither to the task of lighting up India nor to powering its industrial development. The stated objectives of

TABLE 2.1

Generating capacity, electricity generation, and electricity sales, 1944

		*Installed capacity (kW)**	*Electricity generated (million kWh)*	*Electricity sales (million kWh)*
British provinces	Ajmer-Merwara	1,394	2.969	2.415
	Assam	2,524 (500)	3.888	3.430
	Baluchistan	1,250	2.605	1.929
	Bengal	336,622 (2,360)	845.746	739.126
	Bihar	27,162	73.889	66.656
	Bombay	323,999 (232,114)	1,323.200	1,138.988
	Central Provinces & Berar	20,563	38.670	33.056
	Coorg	74	0.072	0.061
	Delhi	22,285	71.472	61.202
	Madras	135,740 (81,250)	370.338	291.541
	NWFP	10,630 (9,600)	16.495	12.773
	Orissa	1,221	1.445	1.186
	Punjab	89,539 (49,750)	194.737	148.972
	Sind	17,558	37.041	29.787
	United Provinces	144,935 (22,700)	367.346	303.941
Indian states	Baroda	4,761	7.446	5.969
	Bikaner	4,000	11.102	7.759
	Cochin	2,919	5.259	7.381
	Gwalior	3,846	8.553	6.723
	Hyderabad	12,901	28.018	24.044
	Indore	2,980	7.811	6.860
	Jaipur	2,995	5.680	4.449
	Jodhpur	2,600	5.421	4.178
	Kashmir	4,270 (4,005)	20.500	12.273
	Mysore	61,000 (61,000)	280.428	216.570
	Rampur	4,200	7.186	5.692
	Travancore	16,947 (15,400)	70.165	58.311
	Other States	21,069 (340)	33.830	33.791

Source: Office of the Electrical Commissioner, *Public Electricity Supply: All-India Statistics, 1944* (Simla: Government of India Press, 1947). Compiled from data on pages 7–13.

Note: Figures in parentheses indicate hydroelectric capacity.

the 1948 act were to reorient the sector to "provide for the rationalization of the production and supply of electricity, and generally for taking measures conducive to electrical development."[44] Through the act, the assembly members created new public institutions—the Central Electricity Authority and SEBs—that became the nodal agencies for the task of electrification. Although this legislation did not reserve electricity as an entirely public domain—a task accomplished by the Industrial Policy Resolution, 1956—it did set the stage for much broader government involvement in the sector.

During the debates, a few themes were discussed most vigorously. These included the topic of nationalization and questions about how quickly and to what extent the state should take control of the sector. In supporting nationalization, many members were concerned with overcoming existing inequalities in electrification. They contended that the state must involve itself in the sector until sufficient demand, or load centers, emerged in rural areas and small towns, or until incomes rose enough for rural citizens to afford electricity at its cost of supply. Providing electricity to these areas would be unprofitable and no private entity would undertake the investment. Many of those who advocated complete nationalization, which would have meant buying out all extant private power firms, argued that this was in keeping with accepted global practices. One member after another referred to British nationalization and Roosevelt's TVA, the latter, as mentioned earlier, as an explicit model for the Damodar Valley Corporation. Shibban Lal Saxena, from United Provinces, argued:

> We know that England has nationalized its electricity, coal and some other industries. In their Electricity Act they have provided for compensation to present manufacturers . . . I do not think England today is abounding in wealth. She has to keep her life going today with the help of America and yet although the country is in such bad days she has taken over the key industries. I think India is much more solvent than England and she can afford to take over these concerns and pay compensation to the owners of the companies by spreading it over a number of years.[45]

The nationalizers also argued vehemently that private electricity companies disregarded the role that electricity should play in advancing the social good. A. Ayyanger, a member from Madras Province, where 90 percent of generation and most of distribution had already been nationalized, spoke for many in his scathing portrayal of private utilities:

> But what these [private] corporations did was to take away the cream of income from the public and not contribute even a little or a farthing to the expansion of power to the rural areas . . . Corporations were easily established in towns [where] for lighting and other purposes they were charging at the rate of 4 annas per unit whereas under the terms of license they were obliged to supply power for agricultural and industrial purposes at the rate of 9 pies per unit. Therefore, these Corporations always concentrated their efforts only in cities where on account of lighting they would get the largest portion of income, but they tried least to get into the villages and giving [*sic*] power to lift up water from wells and so on.[46]

Opponents of full nationalization argued that India did not have sufficient technological expertise and skilled manpower to fully take over the private sector. At least one member of the Constituent Assembly referred directly to the presence of the Tata industrial group in the lucrative electricity market of Bombay, and declared that it would be unwise for the government to displace their operations.

Ultimately, rather than mandate an entirely publicly owned electric system, the new legislation represented a compromise between the government and private operators. To underscore this point, at the end of the debate, N. V. Gadgil, the minister shepherding the legislation, cautioned that the legislation would disappoint both the socialists and the capitalists: "that it is an experiment in a mixed economy is undoubtedly true."[47] Existing private licenses were to be honored, and state-level agencies were given the discretion to extend licenses when they expired. Following the legislation, some state governments and their new public utilities, such as in Andhra Pradesh and Tamil Nadu, were quite aggressive in nationalizing the sector fully, while others, especially those in the most industrialized states of Gujarat, Maharashtra, and West Bengal, continued to allow private utilities to supply power for decades, including in the large industrial belts around Bombay, Calcutta, Surat, and Ahmedabad that have persisted into the present. In 1951, there were more than 300 private utilities, a mix of mostly distributing and some generating firms, and 270 much smaller state and municipal utilities. Twenty-five years later, there were still forty-nine private and twenty-one municipally controlled utilities, but private firms generated only 14 percent of the total electricity in the country.[48] This included older private utilities, based exclusively in urban areas, and a small number of new licenses granted to rural electricity supply cooperatives.

A second issue that emerged during the assembly debate was the relationship of the new utilities and the provincial governments, a discussion that in many ways anticipated contemporary debates about the electricity sector, as well as public-sector enterprises more generally. Those who were concerned about the dynamic between the SEBs and the state governments sounded much like contemporary critics of state-owned firms. In some regions, like Madras and Mysore (now Tamil Nadu and Karnataka, respectively), electricity was not the purview of a separate public corporation but was rather housed in an Electricity Department within the government itself. Taxes on electricity in these areas had become a key source of provincial revenues. Representatives from these regions therefore opposed the creation of SEBs, insisting that the same work could be done more efficiently within the executive branch.

Those who advocated hiving off electricity into a separate public corporation anticipated the problems of increasing interference by elected leaders. Presciently envisioning a time when electricity would come to be a tool wielded to fashion and sustain political constituencies, these members argued that the SEBs should be given full autonomy. The view of K. Santhanam from Madras Presidency, who was also a member of the Select Committee, exemplifies this sentiment:

> What are [*sic*] the British Government doing? . . . Then again even in America when they wanted to start a national undertaking they established a Tennessee Valley Authority. These democratic governments knew what nationalization meant; they knew that these industrial undertakings should not be left to the vagaries of ministerial change . . . Ministers may change, and changing Ministers may have changing policies; but the day to day administration of industrial undertakings should be continuous and should not be disturbed by political considerations."[49]

Again, an uneasy balance was struck to pacify two opposing camps. The legislation mandated that all of the states would eventually create autonomous corporations, but allowed the states sufficient time to establish these bodies—initially two years from the passage of the 1948 act, but with the promise of more time should it prove necessary. Most state governments took advantage of this provision, many waiting to establish their boards until the late 1950s and even as late as the early 1960s.

As discussed earlier, the issue that had exercised Ambedkar and others during the preindependence conversations about planning was whether and to what extent the central government could insert itself into the electricity sector. The standing committee had further debated this issue while drafting the legislation, but by the time the bill was considered in the Constituent Assembly, this issue was much more muted.[50] In the end, despite the views of Ambedkar, Saha, and others, electricity remained a concurrent subject in which the central government's direct interventions in electrification remained limited to discreet projects like the DVA. Why electricity persisted in the hands of subnational governments is difficult to fully explain. Certainly, there were material interests involved. Taxes on electricity constituted a significant stream of provincial revenue in many regions, particularly in the most industrialized provinces of Bengal, Madras, and Bombay, and in industrialized princely states like Mysore.[51] For example, S. V. Krishnamurthy Rao from Mysore argued:

> The entire personnel is manned by Mysoreans and the entire profits of the department go to the coffers of the State. Nearly one tenth of the income of the State, more than a crore of rupees, is the income that the State gets from the electrical department. The industrial progress of Mysore I may assert is mainly due to the progressive policy that the Government of Mysore has followed in the complete nationalization and a [*sic*] distribution of electricity in the State.[52]

In the same speech, Krishnamurthy Rao further pointed out that the constituent assembly need not look abroad for models of state-led electrification: "We need not go to England or any other country to get an example. Here is a small unit of 29,000 square miles—Mysore State—which has followed this policy [of state ownership] to a successful conclusion."[53]

Certainly, then, provincial and princely India wanted to retain electricity as a decentralized sector. Another piece of the explanation, though, seems likely to be found in the way that earlier institutions imprint on later ones. The initial drafting of the Electricity (Supply) Act of 1948 that was debated in the Constituent Assembly began as early as 1942 and continued through the last years of the Wavell administration.[54] As such, and as one of the members of the drafting committee noted during a committee session, the members were "bound by sec. 126(2) of the Government of India Act which limits the powers of the Central Government to legislating, and not to giving executive directions."[55] The act that framed the rules for one of postcolonial India's largest, most expensive, and most far-reaching economic sectors was, in the end, a product of a liminal political moment. The legislation was enacted just before the Constituent Assembly adopted a democratic constitution for independent India, at a time when British agents had formally departed but while British rules were still in place.

Not only in electricity but in most other jurisdictions as well, the distribution of legislative and administrative subjects between India's two highest levels of government seems to have been resolved in favor of precedent. The Indian Constitution in this regard looks very similar to the Government of India Acts of 1919 and 1935. Advocates of change were largely sidelined. For example, Maulana Azad argued that education should become a central subject to unify the intelligentsia of the country, and Rajkumari Amrit Kaur wanted to transfer health from the provincial to the concurrent list to give New Delhi an equal measure of authority in such a vital sphere.[56] But, as with electricity, both health and education remained as they were in the lists elaborated in 1935.

Critiques of the existing division of labor in the field of development continued after independence. For example, invited by the Indian government to assess the new country's emerging system of public administration, the Ford Foundation's Paul Appleby commented in a 1953 report that "the Centre is without any real powers in almost the entire field of development."[57] In that report, he also advised that even a vehicle like the DVA was still too weak, subject to review by two state governments as well as the center. He advised that development projects that impacted multiple states—which would include many of those in the power sector—should be managed exclusively by the central government with no state involvement.[58] Many other reports produced within India and abroad from the 1950s to the 1970s concurred with this view.

The sectoral characteristics of electricity made the devolution of authority to the subnational states more consequential and perhaps more deleterious than in other spheres, like health or education. In those social sectors, maintaining control at the second tier of government allowed public health and public education systems to be attuned to the particular linguistic, demographic, and cultural needs of the citizens that might differ from one state to the next. In electricity, however, both existing demand for electricity and available inputs (hydro and thermal resources used to generate electricity) varied significantly across India. A larger, more interconnected electricity grid allows load—specifically peak load, or the greatest amount of electricity that is demanded at any one time—to be satisfied with a smaller overall capacity. But since state governments were charged with generating, distributing, and transmitting electricity, each was compelled to develop a fully integrated utility system, ideally large enough to satisfy peak load. The result was a politically convenient but technoeconomically suboptimal utility system, a fact that the central government attempted to overcome through repeated interventions from the 1960s onward.

The Public Electricity System: 1950s–1980s

During two decades beginning in 1950, India developed a robust public utility sector in which the SEBs and a small number of central projects like the DVA emerged as the dominant actors. The proportion of private utilities that continued to exist in a handful of urban zones, as well as the share of nonutilities (a category that includes public and private industries that generate their own electricity, as well as a small fraction of electricity generated by the railways), declined throughout this period (see Table 2.2).

The central government also built a nuclear power sector beginning in the 1950s, but though the nuclear program has consumed vast funds over the years, nuclear energy contributes only a small share to India's total electricity output.[59] The rapidly expanding electric grid consumed on average 18

TABLE 2.2

Electricity produced by public utilities, private utilities, and nonutilities (GWh), 1951–1971

	1951	*1956*	*1960–1961*	*1965–1966*	*1970–1971*
Public utilities	2,458 (33)	5,294 (45)	11,016 (55)	26,072 (71)	49,561 (81)
Private utilities	3,400 (45)	4,368 (37)	5,921 (29)	6,918 (19)	6,267 (10)
Nonutilities	1,656 (22)	2,210 (19)	3,186 (16)	3,835 (10)	5,384 (9)

Source: Data from P. D. Henderson, *India: The Energy Sector* (Washington, DC: World Bank, 1975), 70.
Note: Figures in parentheses are percentages of total electricity produced.

percent of all planned investments under the Planning Commission's five-year and annual development plans, of which the majority was spent by the state governments. For example, in the third plan, state governments dispersed 90 percent of the funds allocated for power (see Table 2.3).[60]

When looking at the sector from the vantage point of New Delhi, two dynamics are at work in India's electricity sector throughout the SEB era, which dates from the 1950s (when these vertically integrated public utilities were formed) through the 1990s (when the momentum shifted in favor of the private sector). The first concerns the allocation of electricity between industrial and manufacturing interests on the one hand, and agricultural and agro-industrial interests on the other. In many states, electricity went from being a development input that favored industrial consumers through preferential pricing and access to one that promoted specific farming communities. The second dynamic concerns the balance of power between the states and the central government in the sector. Once the 1948 act came into force, the central government's primary role was to allocate finances for the sector through the five-year and annual plans. From the late 1960s onward, though, agencies within the central government tried various means over time to surmount or circumvent the established division of labor and insert the center more fully into electricity.

Electrifying the Countryside

Almost as soon as the Electricity Act of 1948 was passed, Parliament was pressured to alter the legislation. The demands for amendments came from state governments, the emerging SEBs themselves, and from the Planning Commission and other central government agencies whose reports and committees diagnosed the electricity sector's ills. As if to bear out K. Santhanam's warning during the legislative debates, early amendments eroded the autonomy of the SEBs. An amendment in 1949 permitted the states to

TABLE 2.3

Power in the plans (Rs. crore)

	First plan, 1951–1956	*Second plan, 1956–1961*	*Third plan, 1961–1966*	*Annual plan, 1966–1969*	*Fourth plan, 1969–1974*	*Fifth plan, 1974–1979*	*Annual plan, 1979–1980*	*Sixth plan, 1980–1985*	*Seventh plan, 1985–1990*
Power sector spending	385.4 (19.7)	452.0 (9.7)	1,252.3 (14.6)	1,212.5 (18.3)	2,931.7 (18.6)	7,399.5 (18.8)	2,240.5 (18.4)	18,298.6 (16.7)	37,895.3 (17.3)
Total plan spending	1,960.0	4,671.8	8,576.5	6,625.4	15,778.8	39,426.2	12,176.5	109,291.7	218,729.6

Source: Planning Commission, *Indian Planning Experience: A Statistical Profile* (New Delhi: Government of India, 2001), 31–33.
Note: Figures in parentheses are percentages of total plan expenditure.

appoint their own chief engineers and other members of the government to become chairmen and members of the SEBs. A 1956 amendment ensured that the SEBs would take "policy directives" from the state government and seek government approval for any project over Rs. 1.5 million, which included almost everything in this capital-intensive sector.[61] The same amendment also retracted the ability of the SEBs to set tariff levels independently; instead, the boards would have to secure government sanction.

One reason that state governments sought to impose greater oversight over the functioning of the public utilities was increasing dissatisfaction with how the utilities allocated and charged for electricity to different categories of consumers. India's electricity pricing structure today tilts heavily in the direction of giving farmers cheap or even free power and charging higher rates to industrial consumers. But in the first decades after the SEBs were formed, price distortions skewed in the opposite direction. Even though the rationale for nationalizing the sector was to privilege underserved rural constituencies, in most states during the 1950s, tariffs for rural consumers were higher than those for industrial consumers. The SEBs justified their pricing matrix the same way that private utilities had earlier, by pointing to how much it cost to supply power to different consumers: it was far more expensive for SEBs to supply electricity to remote rural areas with low overall demand, whereas it was much less costly to supply industrial consumers who tended to be geographically concentrated and consumed electricity in large quantities. But the tariffs that most SEBs charged industrial consumers were not even on par with the costs of supplying them with power. A 1962 World Bank mission to India reported that in most states in India, industry, which used the lion's share of all electricity, was charged on average far less per unit of electricity than the marginal cost of production.[62] The Bank recommended that SEBs substantially increase industrial tariffs, arguing that although low tariffs for industrial consumers might have had a positive psychological effect in the past, they were no longer serving to stimulate private investment. Instead, the report's authors argued that private capital was discouraged by a lack of available power, which itself was a function of an electric system that could not expand because of inadequate revenues.

Likewise, an early Planning Commission study of electricity pricing reported that rates for industrialists in many states were not adequate to cover the cost of producing electricity. The reasons for this again were located in past practices. When electricity was first introduced to urban and industrial areas, utilities often charged concessional rates to attract consumers, particularly appealing to the largest industrial customers who would use the largest quantities of energy. Residential consumers were then charged higher rates to compensate, a practice that continued into the era of SEBs.[63] In the mid-1960s, industrial users of electricity were the largest category of consumers and paid the lowest cost per unit of electricity. One way to gauge the extent of subsidies is to compare the percentage of total units sold to a particular category of consumers and the value of sales from the same category (see Table 2.4):

Whereas industrial consumers accounted for close to three-quarters of all the electricity consumed, sales to industry made up only half of total revenue from electricity sales. Residential consumers were skewed in the opposite direction; they accounted for just less than 8 percent of all electricity consumed, but sales to them amounted to more than 22 percent of total sales.

One consequence of the subsidized rates for industrial consumers was the consistent inability of most SEBs to finance their expansion through internally generated surpluses. The central government appointed a committee in 1964—commonly known as the Venkataraman Committee after its chairman, R. Venkataraman—to investigate the financial aspects of the pricing system and recommend amendments to the 1948 act. Already by that point, the SEBs were deemed to have deep financial problems. In 1961–62, only a handful of boards could repay even interest charges on government loans. Among those, which included Mysore, Madras, Madhya Pradesh, and Uttar Pradesh, were also the boards with the most extensive rural electrification, including Madras and Andhra Pradesh (see Table 2.5). The problem, then, was not in the high costs of rural electrification alone but in the

TABLE 2.4

Pattern of electricity sales (1965–1966)

Category of consumer	*Units sold (million kW)*	*Percentage*	*Sales value (Rs. million)*	*Percentage*	*Rate per kW (paise)*
Domestic and residential	2,335	7.8	542.2	22.7	23.0
Commercial	1,650	5.4	286.9	12.1	17.4
Industrial	22,409	73.8	1,193.2	50.1	5.4
Waterworks and sewage pumping	625	2.1	46.7	2.0	7.5
Traction	1,154	3.8	54.9	2.3	4.8
Public lighting	280	0.9	59.3	2.5	21.2
Irrigation	1,892	6.0	198.5	8.4	10.1

Source: Central Electricity Authority, *Public Electricity Supply, All-India Statistics (1965–66)*, cited in K. Venkataraman, *Power Development in India*, 69.

TABLE 2.5

Rural electrification as of 1963

State	*Total number of villages and towns according to 1951 census*	*Average population per village*	*Number of villages and towns electrified as of March 1963*	*Percentage of villages and towns electrified*	*Percentage of electricity sold for agricultural use, 1962–1963*★
Andhra Pradesh	26,743	1,345	3,516	13.1	12.1
Assam	25,355	468	69	0.3	Not available
Bihar	68,078	682	3,193	4.7	3.0
Gujarat	18,463	1,117	1,119	6.1	7.8
Karnataka	26,167	899	3,183	12.2	3.4
Kerala	4,685	3,602	1,716	36.6	2.9
Madhya Pradesh	70,236	461	713	1.0	0.7
Maharashtra	36,441	1,084	1,731	4.8	1.8
Orissa	48,437	363	331	0.7	Not available
Punjab	21,636	964	4,742	21.9	4.0
Rajasthan	31,931	631	336	1.1	6.6
Tamil Nadu	18,646	1,804	7,762	41.6	22.0
Uttar Pradesh	112,208	658	5,628	5.0	Not available
West Bengal	38,591	906	646	1.7	0.6

Sources: Planning Commission, *Report on Evaluation of the Rural Electrification Programme*, 19.
★ Percentages of agricultural sales are from Venkataraman Committee Report, 19.

continuing subsidies to the high-tension industrial consumers. The committee noted that some state governments used subsidized electricity to advance particular economic goals and compete with other states for investments.[64] The Venkataraman Committee recommended amendments to the 1948 act that required SEBs to generate a 3 percent return on the value of the SEB's fixed assets, and that SEBs should increase tariffs to reflect the cost of supply for industrial consumers.

Aside from advising SEBs to charge higher rates to industrial consumers, the central government's more direct intervention in the electricity sector during the 1960s was a renewed attention to rural electrification. Electrifying rural India was an important objective for early planners, and yet the 1948 act left the manner in which this was to be accomplished—the speed and method of finance—to the discretion of individual states.[65] Accordingly, there were wide discrepancies in the way the states undertook the task of rural electrification over the ensuing decades.[66] Whether and how quickly individual states took up the challenge of rural electrification depended on whether and how agrarian interests were prioritized in state-level development agendas.

In the mid-1960s, India experienced several years of acute food shortages that compelled a renewed focus on agrarian productivity in the *Fourth Five-Year Plan*. Accompanying this was a marked shift in the goals of rural electrification from village lighting to pump set energizing, as the productivity-related benefits of electrification came to seem more important in this period than its capacity to foment social change. A mammoth report produced by the second all-India rural credit review committee in 1969 emphasized the importance of providing additional financial means to the SEBs to offset the losses from the costly task of rural electrification. While new rural credit facilities were already providing avenues for farmers to purchase pumps and dig wells, the expansion of the electric grid was proving to be a bottleneck in agricultural modernization. The committee therefore recommended setting up a new financial corporation for rural electrification that would extend low-interest loans to SEBs.[67] In 1969, prompted by what was perceived to be a disappointingly slow and uneven pace of a rural electrification in most states, the Indian government formed the Rural Electrification Corporation (REC), with US bilateral aid, to finance transmission extensions to underserved rural areas.

The use of electrified irrigation pumps to expand the access to groundwater irrigation and to lower its costs first dates to the 1930s, when British provincial administrations sponsored tube wells and lift irrigation facilities in the districts of present-day Bihar, Uttar Pradesh, and Punjab. Madras Province also had a robust rural electrification program and the highest rural electricity sales at independence. In the decades following, only a few other states began to exploit this means to increase agricultural productivity. While some state governments provided explicit policy directives to their SEBs and backed these up with finance subsidies to promote rural electrification, most did not. The data for village electrification and agricultural sales from the early 1960s bear out the great variation in the rural grid (see Table 2.5).

So the necessity for targeted financing such as that provided by the REC emerged from the observations that rural electrification was inadequate in most regions, uneven across states, and that in some cases, SEBs repurposed funds intended of rural electrification to finance new generating capacity. The financial capacity of states to promote rural electrification was likewise uneven. In Maharashtra, West Bengal, and Gujarat in particular—three industrially dynamic states with large generating capacity and high rates of industrial consumption—the industrial areas were all served by private companies. The capacity of their SEBs to generate resources to finance rural electrification was limited compared to the industrially advanced states of Tamil Nadu or Karnataka, both of which had publicly owned sectors. Tamil Nadu had evolved a rationale and a manner to subsidize rural electrification much earlier than elsewhere in India, which is perhaps best demonstrated by the significantly higher rates of both village electrification and electricity sold for agricultural uses.[68]

While supplementing the total quantum of rural electrification financing, the central government's REC also tried to directly intervene in the social aspect of electricity access. All of the states that received REC funds were required to electrify what were called *harijan bastis* (village areas inhabited by former untouchables) along with their overall village and farm electrification. Likewise, the REC financing attempted to correct for intrastate regional imbalances in electrification by charging lower interest rates and extending the repayment period for electrification proposals in underserved regions.

Generating Power in the Center: NTPC and NHPC

Central government agencies and departments had for some time issued reports that advocated increasing the central government's role in the electricity sector, echoing Ambedkar's concerns from the colonial-era planning exercises and the Ford Foundation's early assessment of Indian public administration. By the early 1970s, technological changes had tilted the sector in favor of even greater economies of scale. Advocates of a more centralized sector could point to advances in high-voltage transmission technology that ensured that less electricity was lost in transit, and to new technologies in large-scale thermal generation. An optimal model for an electricity network now consisted of large thermal generating plants sited close to coal fields, and high-voltage transmission lines that carried the electricity over large distances to centers of demand. This was a very different model to what then existed, in which each SEB was actively trying to build up its own power plants, whether there were nearby coal deposits or whether coal had to be transported at great cost from neighboring regions. Given this changed technological terrain and the eroding financial and technical health of the SEBs, both the Indian central government and international funding agencies were therefore increasingly keen to adopt national rather than state-level planning and development for the sector.

The Power Economy Committee of the Ministry of Irrigation and Power produced a series of reports in the early 1970s, proposing that "since 1965, a great deal has happened to reinforce the conclusion that the State is now too small a unit for satisfactory planning and operation in electricity . . . the time has come when the Centre should play a vital role."[69] But when the committee's model for centralization was broached at a conference of state power ministers in New Delhi in 1971, the representatives of a handful of states with the largest and most established utilities—including Maharashtra, Karnataka, and Uttar Pradesh—protested that taking electricity out of the hands of the SEBs and state governments would deprive them of a crucial tool to influence the pattern of development in their regions as well as a crucial source of revenue.[70]

The World Bank's assessments of what ailed India's power sector further noted that state-level electrification had led to inefficient use of capital. A 1975 report argued that "in the identification of projects there is a bias against major schemes which extend beyond State boundaries,"[71] and cited this as one of the most important obstructions to progress in India's electricity sector.[72] According to the Bank, the solution was to further centralize the sector through several means. The simplest and most easily executed solution was to strengthen the hand of the central government through institutions that already existed on paper. The 1948 act that had created the SEBs had also created the CEA, which was officially formed in 1950 but up to this point had only a thin administrative existence with no real staff or clear mandates. Now, the Bank advocated that the CEA should be put in charge of overall planning and development as well as coordination among SEBs; to accomplish this, the government amended EA48 in 1976 to expand the scope of the CEA's authority.

Besides bolstering the CEA, the Bank proposed three potential methods to centralize the power sector. A radical option would be to assign all electricity generation and bulk supply to the central government and leave only distribution to the states. A less radical option would be to reserve all new generating plants for the central government but allow the vertically integrated SEBs to continue as they were. A third, much more modest proposal would be to keep everything in place but reserve all new large capacity thermal and hydroelectric generation for the central government. Since the first would "raise basic political and constitutional issues," the third was deemed less technically and financially optimal but "more acceptable."[73] New Delhi followed the third option in 1975 by establishing the National Thermal Power Corporation (NTPC) and the National Hydroelectric Power Corporation (NHPC) to expand electricity generation.

Alongside its efforts to centralize the sector, during the *Fourth Five-Year* plan, New Delhi began to regionalize the sector into clusters of contiguous states. The creation of regional grids would allow SEBs to coordinate technically with each other (for example, to ensure that the grids operated within the same frequency) and to respond quickly to episodic crises of faulty transformers and grid collapse in neighboring states. States with excess generating capacity were already selling electricity to those with insufficient power, but these sales negotiations frequently erupted in conflicts over pricing that the central government tried to resolve, often unsuccessfully. The creation of regional boards was also expected to constitute new forums for dialogue and conflict resolution in these cases.

The five regional zones—one each in the northern, southern, eastern, western, and northeastern parts of the country—functioned very differently. By 1975, only one—the southern grid—was working as it was intended to by facilitating power flows across state lines and thus enhancing the overall efficiency of the southern SEBs. An interconnected electricity grid created the possibility for shortages in one state to be met with excess capacity in another.[74] An additional benefit was seen in the higher average plant load factor in the south as opposed to other regions; by sharing power between regions and sending electricity to where it was necessary, these integrated grids were able to make better use of their generating capacity. The central government's Powergrid Corporation, formed in 1989, took over the regional grids starting in 1994 in an effort to create a single national electric grid, a task that is still under way.

By subjecting the tariff-setting process to state government approval, early amendments to the 1948 act had created conditions for state governments and their utilities to depart from earlier pricing and allocation logics, in particular dismantling the preferential treatment for industrial users of electricity. The amendments that eroded SEB autonomy opened the door for political and electoral considerations to influence patterns of electrification, and the tariff-setting process in particular. But, as I show in the state-level analyses in the following chapters, the rate and direction of change varied from state to state as a function of the political strength and organization of agrarian constituencies. In many states, powerful new farmers' movements demanded increased support for agricultural inputs—particularly subsidized electricity for irrigation—and, in some cases, they were wildly successful in achieving their goals.[75] Beginning in the 1970s, one state after another responded by subsidizing electricity, a policy that was also justified because of the precedent set by subsidized canal irrigation rates and the new priority given to increasing agricultural production beginning in the late 1960s.[76] For states with robust rural electrification programs, the consequences were enormous. Productivity, cropping patterns, and hydrology were all affected. Perhaps the most negative consequences of breakneck rural electrification were the ecological crises of declining groundwater levels and depleted aquifers.[77]

Also during the 1970s, some states switched from metering agricultural consumption to flat-rate billing, a seemingly minor change that was intended to provide administrative efficiency but had a cascade of unintended negative consequences.[78] The lack of metering allowed the SEBs to hide transmission and distribution losses and the theft of power under the category of agricultural consumption. To offset the losses from lower tariffs for agriculturalists and concealed theft, the SEBs gradually began charging higher rates for industrialists, resulting in a system of cross-subsidies. In many states, the expansion of electricity to ever more consumers was not matched by adequate investments in the transmission grid, which reduced the quality and reliability of network electricity. By the 1980s, many parts of India were gripped by episodic and increasingly severe power crises. As a response to the increasingly unstable electricity grid and higher prices, many industrialists began to abandon unreliable and costly grid power in favor of in-house captive generation, leaving the SEBs with fewer lucrative customers.[79]

All of these tendencies further accentuated the financial difficulties that many states' electricity sectors already faced by the end of the 1980s and, in turn, the overall fiscal condition of state governments. Moreover, in a recursive cycle, agricultural subsidies had the political effect of further strengthening rural interests within the states, which in turn made the subsidy regime more difficult to reverse. These tendencies were more apparent in some states and entirely absent in others. It was only in the latter cases that the publicly owned utility system could be dismantled, as we see in Chapter 4 on Odisha. The performance of the SEBs also deteriorated due to corruption, which was famously difficult to measure.[80] Some of the activities described as corruption were not so much the product of the internal operations of the utilities as of the relationship between elected officials and the appointed administration of the boards. For example, the management of the SEBs was subject to political interference that ranged from personnel decisions to the awarding of contracts for construction and manufacturing in the sector.[81] These myriad financial, political, and technical problems in the electricity sector combined in the late 1980s with new ideas about ownership and control of utilities. The result was to set in motion fundamental changes to the sector in New Delhi.

Market Reforms in the 1990s

In 1991, the central government announced that the Indian electricity sector would no longer be configured exclusively in the public sector, justifying this position with reference to the managerial and fiscal failures of the SEBs, private industry's dissatisfaction with public power, and India's balance of payments crisis.[82] Over the following decade, state and central governments began to reorganize the Indian sector around the market, encouraged by a new global electricity paradigm that favored private ownership and competition over publicly owned monopolies.[83] In the first phase of market reforms, the central government encouraged private investment in generation. In the second phase, and largely in response to inadequacies of the first, the central government encouraged state governments to dismantle their SEBs, privatize distribution functions, and create independent regulators. Finally, in 2003, the central government passed new electricity legislation that replaced the Electricity (Supply) Act of 1948, and gave a new architecture to the sector.

Phase One: Independent Power Producers

In 1991, while faced with a balance-of-payments crisis, the central government amended the Electricity (Supply) Act of 1948 to make the sector more attractive for private investment. Introducing the debate in the lower house of Parliament, Minister of State Kalp Nath Rai argued that the amendment was necessitated by a paralyzing scarcity of financing for the sector.[84] The amendment had bipartisan support. Speaking for the largest opposition party, the Bharatiya Janata Party (BJP), parliamentarian Vasundhara Raje remarked that electricity is a sector "where the States have failed to fulfill their responsibilities . . . The BJP has consistently been advocating the cause of liberalization, deregulation and privatisation, wherever necessary."[85] Even those ideologically opposed to privatization echoed this argument, albeit reluctantly. Congress politician Sriballav Panigrahi from Deogarh in Odisha began, "It is no pleasure to invite private sector to this area. But there is compulsion," and then went on to describe the recurring power shortages that were forcing the government to take this step.[86] Remarkably, the issues that were so passionately debated in the 1940s—regional balance, urban-rural balance, and electricity as a tool to unite and modernize India—were almost entirely absent. The only opposition to the amendment came from Communist members, and even they ignored broader issues about who gets electricity and at what price, focusing instead on how best to protect workers in the public sector. In 1991, the concerns of policy makers with regard to electricity were far removed from those in 1948; the idea of electricity as an instrument of social development and national unification no longer aroused political passions.

That the electricity debate took place under the shadow of a balance-of-payments crisis is one reason that the 1991 reforms went largely unchallenged outside of Parliament. Also important, however, is the broader context. Industrialists no longer supported government control of the commanding heights and instead embraced the new ideology of reduced state involvement in the sector. Opening up the sector offered industrial firms a way to expand electricity supply and extricate themselves from the financial drain of cross-subsidies. Since many private companies already had developed their own in-house "captive" generation, they were eager to find a way to sell their excess supplies back to the electricity grid.

In 1980, the main problems in the power sector were considered to be technical and organizational: thermal power plants had low productivity; the coal supplies of the SEBs were of poor quality and unreliable; and the SEBs needed better management.[87] Within a decade, though, the fiscal losses of the SEBs had grown substantially and became the new focus of critique. The growing quantum of their annual losses contributed substantially to the poor fiscal position of state governments.[88] By the early 1990s, many SEBs were locked into providing highly subsidized electricity to agricultural and residential consumers, and they were simultaneously prevented from raising tariffs elsewhere in the system. While the price of coal and the cost of transporting it to generating stations increased, SEB revenues remained unchanged. As a result, power utilities were increasingly unable to even repay loans, leave aside finance new capital investments. Because coal, railways, and also the power generation capacity of the NTPC and the NHPC belonged to the central government, the inability of these enterprises to collect payments for coal, transport, and energy from the SEBs became a source of tension between the center and the states, and a further reason for the center to demand a change in the status quo. The states also came under pressure from international financial institutions, which demanded reforms in the SEBs before granting new loans.[89]

In addition, the inadequacy of power sector data contributed to the unalloyed momentum in favor of private generation. Transmission and distribution (T&D) losses by SEBs—including theft—were consistently under-reported, while agricultural consumption was inflated. By not metering farm consumption of electricity, and instead only estimating it based on the size of the irrigation pump set and other equipment, the SEBs had created for themselves an elastic category of consumption in which they could hide increasing T&D losses as well as theft. This erroneous reporting contributed to the view that the problem was scarcity alone, rather than a combination of scarcity, technical inefficiencies, and theft. Since the early 1990s, when the SEBs began to fully record and report agricultural consumption, one SEB after another upwardly revised its figures for T&D losses, suggesting that capacity addition alone could never have solved the sector's difficulties.[90] At the time, however, there was a near consensus that a power deficit was the main problem.

With so much momentum in favor of the 1991 amendment, there were few countervailing pressures opposing it. There was not yet a broad-based opposition to market reform, and following on the heels of the economic crisis of mid-1991, the public was prepared to support policy changes that would avoid future crises. Although the public utilities constitute one of India's largest employers, labor unions in the sector had not yet organized effectively to oppose privatization.[91] Indeed, the policy on Independent Power Producers (IPPs) was not initially perceived to threaten labor or public utilities, so there was not a strong incentive for labor opposition. Similarly, there was no broad agitation from farmers against the reforms because the IPP policy was not discussed in the context of eliminating subsidies, which continued and even expanded throughout the 1990s.

The central government was confident that by creating an environment conducive to private capital, the electricity sector would rapidly become a popular destination for new investment. The annual report from New Delhi's Ministry of Power for 1991–92 optimistically described the new policy environment this way: "The scheme forumulated [*sic*] under the policy throws the electricity generation, supply and distribution field wide open to private enterpreneurs [*sic*], opening up profitable investment opportunities. That offers a package of incentives which investors, both from India and overseas, will find really attractive."[92] Indeed, the incentives to private power investors were substantial. The initial licensing period for private generators was extended from twenty to thirty years, and the subsequent renewal period was increased from ten to twenty years. Similarly, the rate of return on capital investments, formerly set at 2 percent over the Reserve Bank of India rate, was increased to 5 percent on all investments made after the legislation came into effect. Foreign equity participation was fully liberalized, and an expedited single-stop approval process was created to replace the multiministry approval process of the past.[93]

After the 1991 amendment was passed, private firms rushed to sign Memoranda of Understanding (MOUs) with central and state governments. In 1992, the central government assigned "fast-track clearance" to eight IPP projects, which allowed them to leap over licensing hurdles in order to expeditiously address the capacity shortage. In just five years' time, by 1996, the government had received 190 proposals from private companies, which, if completed, would have produced more than 75,000 MW of electricity. Only fifteen went on to the stage of applying for a technoeconomic clearance from the CEA.[94] MOUs were signed for plant construction all over the country. Of these, one of the largest and most capital-intensive, and the one that would have resulted in the highest-cost power, was signed by Enron and the Maharashtra State Electricity Board (MSEB) in 1992 to construct a gas-fired, 2,000 MW power plant in Dabhol, in coastal Maharashtra.

From the start, the wisdom of Enron's Dabhol project had been widely questioned. Both state governments and the Enron Corporation were accused of acting with a lack of transparency and regard for the public interest.[95] A range of actors from the CEA and the World Bank to social movement activists questioned the economics of the project, pointing out that the high projected cost of power and the dollar denomination of the contract would expose the power purchaser, the Maharashtra SEB, to financial risk.[96] Activists and agrarian communities also protested the environmental and social costs of the project, including inadequate compensation to farmers and potential human rights abuses associated with project construction.[97] The project also became a central issue in state politics, when a coalition of parties in Maharashtra campaigned on an anti-Enron plank, only to cancel and then renegotiate the contract when they came to power. Ultimately, many of the fears about the project's unsuitability proved true, and the Dabhol project landed in a nearly decadelong conflict, which I discuss further in Chapter 3.

Phase Two: Distribution Privatization and Independent Regulation

By the mid-1990s, it was clear that a focus on private investment in generation was an insufficient and possibly counterproductive policy. Not all power purchase agreements with private companies were controversial, nor did all fail as spectacularly as Enron's. Nevertheless, the saga of Enron, ending as it did in a high-profile contractual dispute, suggested the difficulties with expecting IPPs to solve India's electricity crisis. Now the central government reconceptualized the problem to focus on the monopsony structure of the electricity sector.[98] As long as private generating firms were required to sell their power to insolvent SEBs, as the 1948 act mandated, the financial risks to government, and eventually to consumers, would remain high. The central government proposed two solutions to these problems: (1) that the state governments should dismantle their vertically integrated SEBs and privatize electricity distribution, and (2) that each state should establish regulatory institutions to govern tariff policy. Both solutions were attempts to address the problem of political interference with the SEBs, which kept subsidies too high and tariffs too low for SEBs to pay their bills.

The central government's policy recommendations bore an imprint of the World Bank, which in 1993 had rewritten its policy to emphasize private participation in the power sector, independent regulation, and where possible, privatization of existing public companies. In the second half of the 1990s, various Indian states experimented with distribution privatization and regulation. While some states fully privatized distribution, others got only as far as unbundling their SEBs. In large measure, these differences reflect variations in development priorities across the states, a subject that is taken up in each of the state-level analyses in the chapters that follow.

In 1998, the central government passed the Electricity Regulatory Commissions Act. Several states had preempted the central government in creating state-level regulators, or State Electricity Regulatory Commissions (SERCs), the first being Odisha. While restructuring and privatization of distribution utilities had proceeded slowly, and only in a handful of states, almost all states had established regulatory commissions by the early 2000s. The differences in the legal frameworks governing the various SERCs are minimal, but their operations varied from state to state. During their first decade in operation, the main criticisms of the new regulatory regimes focused on their relationship to elected leaders, which some consider far too cozy.[99] The regulators themselves are chosen from among retired or nearly retired bureaucrats, many of whom naturally have preexisting relationships with government from having served in other bureaucratic capacities. Additionally, some state governments have not granted sufficient financial resources to the commissions, prompting the worry that governments will use this leverage to exact politically convenient decisions.

Still others are cautious in holding up regulatory commissions as the solution to the industry's ills because of the possibility of regulatory capture by private companies and utilities.[100] Theoretically, independent regulators are charged with establishing tariffs that balance the competing interests of private companies, utilities, consumer groups, and the public interest. In theory, each of these groups is able to represent its interests before the regulator. In practice, however, private companies and public utilities are better able to organize themselves in such regulatory forums than are consumers. Moreover, there are considerable interstate differences in how regulatory commissions function. One study of the performance of electricity regulatory institutions in India indicates variations in the level of resources and autonomy granted to the SERCs; the degree of transparency and public participation in the regulatory process; regularity in publishing annual reports and holding consultative committee meetings; and the nature of state government and utility interaction with the regulators.[101]

Phase Three: The Electricity Bill of 2003

While reforms in the 1990s were scarcely a runaway success, they did have the merit of attempting to focus on the central problems at hand—distribution reform and SEB losses. With distribution reform firmly on the agenda, the central government reentered the fray with an attempt to replace ad hoc state efforts with an overarching framework to guide reform. The new legislation, passed in 2003, superseded existing legislation in the sector and prepared the ground for a fundamental restructuring of the Indian electricity sector. The intent was to deepen and formalize the transition that was already under way in some parts of the country. To critics of the legislation, by focusing on private participation and relying on market competition to lower prices, the act established conditions for electricity to be managed solely as a commodity rather than as a social good that the state is obligated to provide to its citizens. Many expected the result to be a return to the preindependence conditions that had originally spurred the creation of SEBs, when electricity flowed primarily to cities—areas with concentrated loads and robust purchasing power.[102] The act, critics argued, entirely neglected rural India and specifically the many parts of the country that were still not electrified.[103] Furthermore, critics on the Left argued that the act failed to deal with a larger structural problem in the electricity sector that affects all utilities in the country. There are multiple actors at both the state and federal levels that must coordinate their activities along the electron production chain. From the dominant fuel source, coal, to transportation by the railways, and finally to the mostly state-owned generating facilities, there are endemic problems of coordination, and electric utilities are the final repositories for inefficiencies that stream throughout the chain of production. The two most obvious examples of these inefficiencies are from the coal and railways sectors, both of which are owned and operated by the Indian central government. Coal mines were nationalized and put under the authority of Coal India Limited, a centrally owned public-sector company, in 1973. Since that time, theft and corruption have come to pervade coal mining operations. To offset the losses from these inefficiencies, the prices for coal are increased, in turn affecting the finances of state utilities. Likewise, railways are owned by the central government. High freight charges are used to cross-subsidize low rates for civilian rail traffic.[104]

The new Electricity Act was passed by the Indian Parliament in May 2003. In contrast to the debate on the 1948 Electricity Act five decades earlier, the debate in the lower house was minimal, as was parliamentary attendance.[105] The basic thrust of the act was to open the sector to private involvement and scale back the role of state governments. For many advocates of even faster market reform, the act did not do this fast enough or surely enough; the legislation was said to harbor an incumbency bias that protected the interests of existing utilities at the expense of consumers. In a critical editorial about the Electricity Act of 2003, one of its early drafters who had pushed unsuccessfully for including strict guidelines for the introduction of competition in distribution argued that "delay in introducing open access means delaying competition and private investment. Shortages will continue and consumers will have to rely on the public sector, which does not have the resources for meeting the entire demand. This seems a sure recipe for power shortages and high tariffs."[106] Another critical reading of the legislation argued that the numerous rounds of revision to the draft bill resulted in the inclusion of contradictory policy principles. As one news article stated, the act "incorporates viewpoints, ideas, concepts, constraints and requirements from a wide range of interested parties and has thereby undergone a number of changes."[107] The resolution of these contradictions would require several additional policy directives during the following decade.

Conclusion

The preindependence debates about how India's electricity sector should be organized turned on ownership and authority: should the sector be populated by state-owned or private firms? And which level of the state—the central government or the provincial units—should play the larger role? These same two questions recurred throughout the next five decades and were answered by a changing matrix of solutions.

During the colonial period, the sector was regulated by provincial governments but largely owned by private companies. From the 1950s through the 1980s, the sector became almost entirely publicly owned, with massive vertically integrated public utilities organized along state lines. There also remained important private-sector enclaves in a handful of heavily industrial urban regions. Throughout this period, the central government's influence in the sector increased over time, not only through the financial control of the planning and finance commissions but also through more direct involvement in generation and transmission. This trend continued even through the first decade of market reforms, in the 1990s, but to this was added a resurgent emphasis on private investment as the best route to make India energy sufficient. The generating capacity of the central government's corporations (NTPC, NHPC, and DVA) shot up from 36 percent in 1992 to 43 percent in 2002.[108] The transmission network was also strengthened, allowing regions with surplus power to vacate supplies to power-deficit regions. By 2003, the Ministry of Power confidently predicted in its annual report that "by the end of the XIth [five-year] plan, [the coal-producing] eastern and northeastern region shall be the major source of power for the country."[109]

So, by the time the Electricity Act of 2003 was passed, in one sense, the sector more closely approximated early planners' visions of a centrally controlled and coordinated sector. And, yet, because electricity remained a concurrent subject with substantial oversight by state governments, there continued to be important differences in how electricity was allocated across consumer segments and at what price, and the speed and direction of India's market reforms in energy continued to be determined by state governments. In another sense, however, in the increasing role of private actors in what Ambedkar had considered India's new "strategic railway," India's electricity sector had moved further away from the early nationalist vision.

3

Maharashtra and the Politics of Selective Rural Development

Maharashtra, the focus of this chapter, and Odisha, the subject of the next, occupy two opposite sides of the continuum of Indian economic development. This is the case along measures as varied as per capita incomes, indicators of health and education, levels of urbanization, and the structural compositions of the two states' economies. The most recently available figures from the Indian government illustrate well the gulf in socioeconomic development and access to household amenities across states in India, variation that is exemplified by the uneven access to electricity. According to the 2011 Indian Census, 83.9 percent of households in Maharashtra can rely on electricity for household lighting, whereas the corresponding figure in Odisha is just 43 percent.

What explains such stark differentials in development outcomes on India's two coasts after sixty years of centralized development planning and spending that was meant to correct regional imbalances? Colonial-era patterns of investment and other legacies of the past play some role in shaping regionally differentiated endowments in infrastructure and economic productivity. Likewise, the relationship of central governments with those of the states and the patterns of central allocations through the planning and finance commissions affected what individual state governments could accomplish. But within the constraints of historical patterns and the central government's disbursal of revenues, postindependence provincial political regimes also played a role in crafting distinct development priorities for their territories, as the process of electrification and the politics of the energy sector reveal.

In 2005, the central Indian government launched the Rajiv Gandhi Grameen Vidyutikaran Yojana (rural electrification scheme) to inject new funds into rural electrification, a process that had completely stalled in some parts of India, like Odisha, in the wake of utility privatization in the 1990s. The scheme emphasizes rural household access to electricity. But in those parts of India that had attained relatively high levels of village and household electrification prior to this current effort, the bulk of these gains were not the product of a commitment to expanding electricity for the sake of its social benefits. Rather, rural electrification was successful in those subnational units—and Maharashtra exemplifies this tendency—where electricity was conceived as a crucial development input and, as such, became a linchpin of the regional development strategy. Influence over the provincial state by rural political coalitions was an essential requirement for development to prioritize increasing agrarian productivity and agro-industrialization.

Consider again Odisha and Maharashtra as embodying two distinct development strategies. The postcolonial state elite in Odisha inherited a primarily agrarian economy, but sought to capitalize on the state's mineral resources to foment an industrial transformation. In Maharashtra, by contrast, the political elite inherited one of the largest industrial economies in India at independence, located in and around Bombay, but devoted development resources to foster a dynamic agricultural and agro-industrial economic future. Electricity was a key part of this. To reiterate a point made in the introduction, electrification functioned in a recursive sense: rural electrification was born of a commitment to rural and agro-industrial development. In turn, the provision of electricity strengthened the political weight of the rural communities that benefited from it.

In this chapter, I identify several features of postindependence Maharashtra's social and political institutions that help to explain why the provincial state devoted considerable infrastructural resources for rural electrification: (1) a social structure that facilitated rural organizing for development, (2) the emergence of a new set of cooperative institutions for credit and production that could serve as a bridge between state and rural society, and (3) the political ties that bound the Congress Party in Maharashtra to the countryside. These three features operated in a dynamic framework. For example, the cooperative institutions became critical conduits for the flow of development resources to rural areas, helping to bear some of the costs and risks of new technologies like electrified pump sets for farmers who might otherwise have continued with rain-fed agricultural production. The rural caste structure facilitated alliances across groups of farmers who were otherwise differentiated by size of holdings and class position, and caste solidarities were in turn strengthened through the working of the cooperative institutions. Finally, the Congress Party recruited local party workers and party candidates from rural cooperative institutions, which strengthened the political ties of the Congress-led government to the countryside.

As discussed in Chapter 4, in Odisha, the low level of rural access to goods like electricity has continued from the early decades of state building into the contemporary period. Odisha's material and financial resources were put instead toward industrial efforts. During the colonial period in Maharashtra, too, access to electricity was concentrated almost exclusively in the urban centers of Bombay Province. But following independence, the postcolonial state government worked quickly to redistribute resources toward the countryside. Particularly evident from the time of the state's formation in 1960 through the 1980s, the rates of access and levels of consumption of electricity increased more quickly for farmers than any other category of consumer in the state. As pointed out in later sections of this chapter, however, the benefits were not evenly distributed across rural Maharashtra. Rather, the penetration of electricity was concentrated in the rich farming districts to the east of the Western Ghats, or Western Maharashtra according to the subregional nomenclature. There, electrified tube wells opened up a crucial new source of irrigation that supported the emergence of one of Maharashtra's fastest-growing new agro-industries: sugarcane farming and sugar production. From the 1960s through the 1990s, Maharashtra's "sugar barons," as they became known, were the beneficiaries of a steady flow of infrastructural resources in the form of rural electrification and other input subsidies. I analyze this process of selective rural developmentalism by looking at how electricity resources helped to forge the state's postindependence sugar industry. Although sugar factories and sugarcane fields are found throughout the state, the highest concentrations of both are in the districts of Western Maharashtra.[1] Household electricity access followed from this, and although Maharashtra is today one of the wealthiest states in India, household electricity use continues to be spatially unequal (see Map 3.1).

Despite the economic weight of one of India's largest metropolises, Mumbai, the political center of gravity in Maharashtra has rested in the countryside for several decades since its formation in 1960. Unlike in other parts of India, the rural landed communities in Maharashtra had staked an early claim to political power beginning in the decades before independence, and consolidated this hold soon after the state was formed in 1960. If the initial condition for such a rural developmentalism in one of postindependence

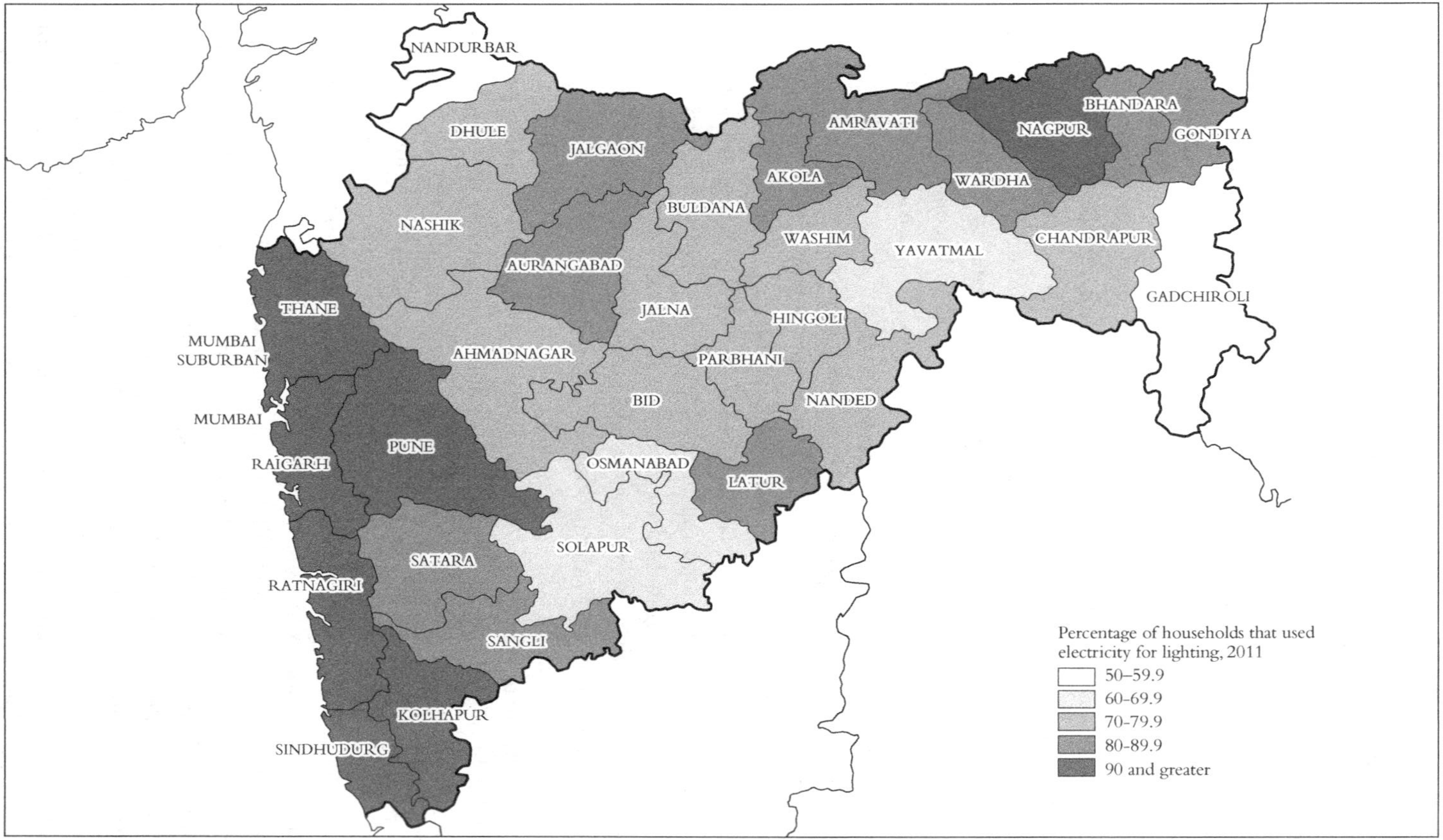

MAP 3.1 *District household electricity use for lighting in Maharashtra, 2011*
SOURCE: *Census of India, 2011.*

India's most industrially advanced states was the presence of political elites with deep ties to the countryside, the process of channeling state resources to the countryside only further entrenched the political clout of rural Maharashtra.

According to one interpretation, politics in Maharashtra represents a tacit understanding between the economic power in the state, concentrated in Mumbai, and political power, held by a rural elite.[2] Another scholar has observed that while the "expansive elite" in the countryside is a powerful force in the state's politics, their role is "more reactive than initiatory."[3] What the rural elite exercises, according to this view, is a significant veto power over policies detrimental to their interests. Reading the state's electricity system as a metaphor gives us a different image of the state's political economy. Rather than a neat bifurcation between political and economic power, or a merely reactive rural sector, we instead see how two distinct political economic logics are held in balance through a careful structuring of institutions. Prior to the privatization drive of the 1990s, Maharashtra was one of the few states in India with a mixed public-private structure of ownership. Private utilities have long served the most lucrative market, in Mumbai, outlasting even the wave of nationalization that took over most private utilities from the 1950s through the 1970s. The rest of the state was served by a large, vertically integrated public utility—the Maharashtra State Electricity Board (MSEB)—that devoted vast resources to rural electrification from the 1960s through the 1980s, and continues to subsidize prices for rural consumers. Maharashtra, then, has two parallel systems, one privately owned and serving the state's largest industrial and urban constituencies, and the other a publicly owned system that transfers public resources to the state's most influential rural constituencies.

The last section of the chapter looks at how the carefully tended relations between the agro-industrial interests and the provincial government from the 1960s through the 1980s structured the kinds of policies that the state government chose to pursue as a part of India's general turn to the market in the 1990s. Although Maharashtra was considered an early candidate for utility privatization, it bypassed the model of utility privatization practiced by Odisha. One reason for this is the role of electricity in the matrix of the state's agricultural subsidies. Since the 1960s, these subsidies—for electricity, fertilizers, and prices—had become increasingly important as the glue that bound the state government to the larger commercial farmers and agro-industrialists, particularly in the western districts of the state. Instead of privatizing these utilities outright, which would have threatened the system of cross-subsidies on which Maharashtra's farmers relied, the government of Maharashtra experimented with alternative ways to augment energy supplies in the state. I highlight two policy innovations in Maharashtra's energy sector that allowed the state to increase electricity supplies without dismantling the existing structure of the sector. The first, Enron Corporation's ill-fated power plant in Maharashtra, became one the most notorious projects in market reforming India and a parable for how *not* to structure foreign investment. The second was a policy to encourage biofuels. Among India's states, Maharashtra was one of the most receptive to these alternative energy sources. Promoting electricity generation based on bagasse (sugarcane pulp) became a way for the government to further support sugar cooperatives, which by the 1990s had begun to decline in efficiency and productivity.

From Bombay Province to Maharashtra State: Postindependence State Formation

Maharashtra occupies a portion of the west coast of India, continuing far inland into what is known as the Deccan Plateau. The state is one of the largest in terms of both territory and population, and by most measures, it is also among India's most developed and wealthy states. The capital city of Mumbai (formerly Bombay)[4] and its environs host important financial, entertainment, real estate, and manufacturing sectors. The remainder of the state is made up of several subregions, each with a distinct economic base, resource profile, and history. These differences among the regions have affected the nature and distribution of political power and resource allocation in the state in important ways.

Prior to 1960, present-day Maharashtra and Gujarat were governed together in a composite region called Bombay Province, the boundaries of which were largely inherited from British administration. Following independence, social movements across India demanded that states be reorganized following linguistic borders. In 1960, the efforts of one such movement in the Marathi-speaking territories, the *Samyukta Maharashtra* (United Maharashtra) movement, succeeded in splitting the territory into two states, with metropolitan Bombay serving as the state capital of Maharashtra.

The Samyukta movement went through several periods of resurgent organization and quiescence from the 1930s to 1960, when its goal was finally achieved in the breakup of Bombay State. One period of heightened activity occurred during the 1950s, when the national government was in the midst of considering the extent to which language and linguistic identities would be adopted as the ordering logic of India's internal boundaries. In the petition submitted to the States Reorganization Commission in 1954, the Samyukta movement couched its demands in terms of cultural autonomy. As the authors of the submission note, "commonness in language denotes over most areas of the country membership of a regional society and possession of a distinctive cultural, literary and historical tradition."[5] Indeed, the close affinities of the demand for political autonomy with the sphere of cultural politics is evident from the early history of the Samyukta movement, which grew out of literary conferences (*Maharashtra Sahitya Sammelan*) that were held throughout Marathi-speaking British and princely India beginning in the late 1880s.[6]

The movement's success hinged on emotionally resonant appeals to the cultural solidarities of Marathi speakers, but one can discern a set of interrelated political economic concerns that operated alongside these cultural claims. Consider, as an example, the views of D. R. Gadgil, a prominent intellectual, economist, and policy maker both within New Delhi and within Bombay State and later Maharashtra. The 1954 submission, which Gadgil helped to draft, made a case for why approximating cultural homogeneity in India's constituent units was an important precondition to achieve the twin goals of economic development and social democracy. As the authors note: "With all areas waiting for development and clamouring for greater expenditure, rival claims are always difficult to reconcile . . . [with] each language group in a multi-lingual state considering that it was being unfairly treated."[7] Reorganizing India's internal boundaries according to language, then, would eliminate these opportunities for economic conflict between linguistics communities. Although the official submission to the States Reorganization Commission was agnostic about whether these accusations of unfair treatment had any basis in fact, Gadgil had separately expressed concern about precisely this kind of spatially uneven development expenditure in correspondence to Nehru.[8] In particular, Gadgil singled out the way that funding for rural electrification favored the districts of present-day Gujarat and bypassed the Marathi-speaking areas.

B. R. Ambedkar, who was one of the drafters of the Constitution and had taken a leading role in planning for the energy sector in the years prior to independence, also expressed fears about the unequal allocation of resources across regions in the combined Bombay State. As discussed in Chapter 2, during debates and speeches about planning and development prior to independence, Ambedkar had expressed fears about the dangers of parochialism and argued that more centralized energy, transportation, and communications systems would be the fastest, more efficient route to industrialization. But by 1955, when the federal structure of electricity was already entrenched, Ambedkar wrote in favor of linguistically organized federalism.[9] Armed with data that showed persistent inequities in the way that political power and economic resources were allocated between the two major linguistic regions of Bombay—the mainly Gujarati-speaking north and west and the Marathi-speaking south and east—Ambedkar advocated dividing the state along linguistic lines.[10] Within Bombay Province, the per capita development expenditures (in rupees) during 1950–51, 1951–52, and 1952–53 for the Marathi-speaking districts were 1.7, 2.3, and 1.8, and for the Gujarati-speaking districts, they were 2.9, 3.1, and 3.2. This discrepancy prompted Ambedkar to exclaim dramatically, "What an injustice? Can anybody blame the Maharashtrians if they felt disgusted with the mixed State of Bombay"?[11]

The contemporary state that was the outcome of the language movement has a mix of regions that, while united by language, have distinct climates, economies, and political histories. The three primary regions of the state are denoted as Western Maharashtra, which includes the city of Mumbai, the Konkan littoral, and the Sahyadri mountain range (also known as the Western Ghats), all formerly part of the Bombay Province under colonialism; Marathwada, formerly part of the Nizam's Hyderabad, a princely state that also included much of present-day Andhra Pradesh; and Vidarbha, composed of former British central provinces and Berar, also once part of Hyderabad. Scattered throughout these three primary regions are other former princely states.[12]

The topography and climate of Maharashtra's regions vary as much as their historical and political antecedents. The coastal strip that runs along the western length of Maharashtra from north to south feels the brunt of the annual southwest monsoon. The quality of the land is inferior to other parts of Western Maharashtra, although there are cash crops like mangoes and coconuts. Separating this strip from the remainder of the state are the Sahyadri Mountains, whose uneven topography is balanced by heavy rainfall and rich soil. The Deccan Plateau comprises the remainder of the state; although the soil is rich, parts of the plateau receive less concentrated rainfall during the monsoon, but get rains during other parts of the year as well. Other parts are arid and easily susceptible to drought. The Marathwada region is one of the poorest in the state. A large part of the land was under the *zamindari* (landlord) system, in which individuals were given the rights to collect taxes for large tracts of land that enabled them to effect a great deal of control over villages in their territory. Most of Western Maharashtra, in contrast, had a *ryotwari* system, a type of land taxation in which revenues were collected from individual cultivators. Differences in land tenure arrangements in colonial and precolonial India continue to be seen in present-day landholding patterns and development outcomes.[13]

Several features of Maharashtra's social composition and political history distinguish it from other parts of India and are critical in the unfolding trajectory of political life in the state. Foremost among these is the nature of caste and class politics in the state, in particular, the role of the region's dominant caste of Marathas. Sociologist M. N. Srinivas wrote that "a caste may be said to be 'dominant' when it preponderates numerically over other castes, and when it also wields preponderant economic and political power. A large and powerful caste group can be more easily dominant if its position in the local caste hierarchy is not too low."[14] This definition applies very neatly to the Marathas, a conglomeration of numerous *jatis*, or "subcastes," that claim a "middle" position in the regional caste hierarchy.[15] Today, the Marathas represent the largest caste grouping in Maharashtra, and from their ranks come the majority of the region's political and landowning elite. Much of the political science and anthropological scholarship on Maharashtra has convincingly shown that the dominance of the Marathas extends across social, economic, and political domains.[16]

Another significant feature of rural caste society in western India is the more porous nature of Maratha caste boundaries relative to other strictly endogamous castes in India. It was not always the case that the term Maratha designated a particular endogamous community; in the nineteenth century and earlier, the term was initially used to refer to the region's Marathi-speaking martial units, and later, a very small group of ninety-six elite families who claimed descent from sixteenth-century warrior chieftains.[17] Over the course of the late nineteenth and early twentieth centuries, however, as part of a conscious strategy by elite Marathas, the caste name came to designate a much larger community of rural peoples that crossed class lines.[18] The British census became a useful tool in the efforts of elite Marathas to strategically affiliate themselves to agrarian communities of lower socioeconomic status. In the years leading up to the 1921 and 1931 census surveys, Maratha leaders exhorted members of the more numerous and socially less influential *kunbi* communities to mark their identity as "Marathas," solidifying their merger to form one large group.[19] The efforts were successful. In both of those two census surveys, the numbers of individuals reporting their identity as "Kunbi" declined, while those reporting "Maratha" increased.[20]

The effects of this political and social maneuvering in the colonial era continue to reverberate in postcolonial Maharashtra's electoral and political fields. According to some estimates, Marathas make up nearly 40 percent of the total population of the state. In many electoral systems, a community of this size would be considered large, particularly if its members voted with one voice. But within India's highly fragmented electorate, the number assumes even greater significance, allowing any party who can speak for the Marathas or even a sizable fraction of them to command a significant plurality.[21]

Unlike in many parts of India, the political field in Maharashtra has been relatively stable, becoming more of a dynamic and multiparty system only from the 1990s onward. Since the state was formed in 1960, the Congress Party endured as the largest party at the state assembly and in Maharashtra's parliamentary delegations, although smaller parties and factional offshoots have circulated this political core. At the core of Congress's endurance in the state was its Maratha base. Whereas non-Brahmin activists in some parts of India created entirely new political organizations, like the Justice Party in what is now Tamil Nadu, in Maharashtra, the newly politicized Marathas instead joined the Congress.[22] After a brief departure of one faction of Maratha leaders to form the Peasants and Workers Party in the 1950s, Maratha politicians returned to the Congress Party in 1960 and firmly held the reins of the party in Maharashtra from then until the mid-1990s. The means by which the Maratha elite retained control of the Congress and, in turn, the reasons that Congress was able to dominate Maharashtra's politics even as the party declined in most other states owes much to the way that state resources were expended to promote certain kinds of rural development.

Before fleshing out this and other connections between the politics of electrification and the larger political economy of Maharashtra State, the next section looks at early electrification in and around the colonial city of Bombay to explain the first half of Maharashtra's parallel development structure. This history of Bombay's private electric system also links to later discussions in this chapter on the various pressures working for and against privatization in the state. Whereas urban industrial and commercial users of electricity might have been expected to favor privatization (as they did in other parts of the country like Delhi), in Maharashtra, the majority of urban users already existed in a privatized system and were effectively neutral in the debates about whether or not the state should privatize.

Bombay's Private Enclave and the Public Utility System

Electricity arrived in Bombay in the 1880s, used first to illuminate public and commercial spaces. One early site was Crawford Market, a major center of commerce in Bombay that today continues to thrive and demand electricity. An exhibition was held in 1882, at the residence of the managing director of Bombay's first distribution utility, Eastern Electric Light and Power Company, to showcase the safety and effectiveness of electricity as a source of domestic lighting. However, several more decades would pass before electricity became popular for household use.[23] Instead, through the end of the nineteenth century, there was irregular lighting in larger markets, on a few major roads, and in municipal buildings, falling into sporadic disuse due to the financial and technical instability of early private generation and supply companies.[24] The government of Bombay issued the "Bombay Electric License, 1905," after which the use of electricity in the city became more regular, albeit, still extremely expensive and therefore sparse.

In the early decades of the twentieth century, electricity generation and usage expanded dramatically. Almost all of this relied on hydroelectric power produced by the Tata Electric Companies, which comprised Tata Hydro-Electric Power Supply (established in 1910), Andhra Valley Power Supply Company (1916), and Tata Power Company (1919). The biggest users of electricity were textile mills, public lighting, and traction in and around the city of Bombay.[25] Though the Tata companies enjoyed privileged political favor, they were not the only electric utility in the city. The Bombay Electric Supply and Transport Company (BEST) had two small thermal generating units that were eventually closed down in 1926, but the main function of the company was to distribute electricity in the city.[26] BEST and the Tatas shared electricity functions, with occasional tension erupting between the two. The agreement finally arrived at by both parties allowed the Tatas to supply electricity from their generating units directly to the largest consumers—factories requiring more than 500,000 units of electricity annually.[27] All other consumers, however, were to be supplied by BEST. BEST, initially registered as both a British company in London and as an Indian company, and later as an Indian company but with a largely British management, operated from 1905 until one week before independence, when it was taken over by the Bombay Municipal Corporation and run as a municipal undertaking. BEST continues to supply electricity in Mumbai. Consumers in suburban Mumbai were supplied electricity by a private utility, Bombay Suburban Electric Supply (BSES), beginning in 1929. From then until the mid-1990s, BSES exclusively purchased power from the Tatas, but began to generate its own electricity in 1995.[28] At independence, aside from these few large, private distribution and generation companies in and around Bombay, there were several dozen small generation and distribution companies located in the region's larger cities and towns. The vast rural regions were not electrified until well into the postindependence period, when the task was taken up by newly created public utilities.

The Bombay Electricity Board was established in 1954 and became the Maharashtra State Electricity Board in 1960. The board's task of rural electrification in Maharashtra was made more difficult by the fact that the largest revenue-generating region in the state, in and around Bombay, remained a separate utility system. Whereas public utilities in other states were able to use profits from well-established and lucrative urban areas to finance more-costly and risky rural electrification, the largest urban area in the state was outside of MSEB's purview. At the time of its creation, there were only a handful of places with a concentrated potential load that the board could electrify. Among these were Akola, Amravati, Ulhasnagar, Kolhapur, Khangaon, Talegaon, and Chand, which were dispersed across the breadth of Maharashtra. Laying transmission lines to connect all of these far-flung cities to the grid would be costly and time-consuming. To counter these financial difficulties, the board determined to take over the private licensees that operated in the larger towns in the state, excluding Bombay. This was part of the MSEB's plan for the period of the *Third Five-Year Plan*, from 1961–62 to 1965–66.[29]

Despite the board's aggressive takeover program, Bombay—the most lucrative area in the state—was left in the hands of private utilities. The existence of private utilities was perceived to be one of the main handicaps of MSEB relative to other public utilities in India. One report pointed out:

> It is necessary to appreciate some important constraints within which [MSEB] has to function and over which it has no control. Its cost structure is not fully comparable with other Boards and even private undertakings in view of the fact that highly concentrated industrial and other loads in the State yielding substantial revenue are not with it. On the other hand, it is solely responsible for [the] relatively less remunerative rural electrification programme. This affects its capacity to raise internal resources.[30]

Even as late as 1971–72, the private and municipal utilities that sold power to the concentrated industrial and residential consumers in and around Bombay accounted for the majority of all electricity sales in the state, equaling 58.8 percent of the total.[31] Of the total sales, 12.8 percent are attributed to BEST, 8.7 percent to BSES, 36.1 percent to Tata Power, and 1.2 percent to a second small, privately owned municipal utility (Thane Electric Supply Company) located in Thane, which was nationalized in 1973.

The largest of Maharashtra's private firms, Tata Power, both generated and distributed power in the state, but this distribution was limited to very large industrial consumers, again the most lucrative ones. To get a sense of the relative size of the Tata consumers, consider that in 1973, although Tata Power sold 36 percent of the electricity in all of Maharashtra, its total customers numbered only 216, or 0.01 percent of all private, municipal, and utility consumers in the state.[32]

Given how advantageous it would have been to MSEB to acquire the customer bases of Tata, BSES, and BEST in the concentrated region of Bombay, it is fair to ask why these companies were not absorbed in the expanding public utility sector when the majority of other private companies throughout the country were nationalized in the 1950s and 1960s. During the debates in the Constituent Assembly, when the merits of nationalization in the electricity sector were being discussed, one member remarked:

> There are powerful vested interests in one or two places which it would be very unwise for us to displace now. Take the Tata interests [in Bombay]; it would be very unwise straightaway to displace it. It would be much better to get a corporation of that nature with its semi-impersonal character to help us in going ahead with our programme of developing electricity rather than tell them, "We are going to take you over," and find ourselves faced with all the difficulty that would ensue in regard to managements and personnel.[33]

The assembly member who made these remarks, T. T. Krishnamachari, was also a member of the Select Committee who met with representatives from private utilities in the process of drafting the 1948 legislation. And yet he suggests that although nationalization made sense as a general policy, stripping assets from a corporation like the Tatas would be contrary to the country's interests.

To what extent Indian capitalists—particularly large industrial capital—supported or undermined India's planned economy is a recurring debate in the literature on Indian political economy. Some scholars emphasize the socialist character of India's political economy, which was established by a developmental state with strong backing from Indian capital. The existence of the *Bombay Plan*, a document published in 1944 and 1945 by a group of India's largest industrialists, including J. R. D. Tata, is taken as evidence that Indian business supported economic planning and public ownership. Chibber counters this, insisting instead that India failed as a developmental state precisely because the Indian business class resisted the creation of a strong state with expanded regulatory functions. For its authors, the *Bombay Plan* was useful as a way to portray themselves as good nationalists at a time when it seemed that the mass protest against colonialism could transform into an anti–private property agitation. Presenting themselves as advocates of planning also allowed them to anticipate and shape the terms of the debate that would follow independence. Far from leading the agenda of nationalization, planning, and state-led industrialization, Chibber argues, Indian capitalists were crucial to the failure of these programs.[34]

Nationalization in the electricity sector, in which India's most lucrative pockets—core industrialized zones—were preserved for private ownership, shows the important allowances that India's model of public ownership made for prominent private firms. In addition to the Tata companies, which generated and sold power to the Bombay region, the private firm Killick Nixon held utility licenses in the other industrial centers of Surat, Ahmedabad, and the Bombay suburbs.[35] The company became a public limited company in 1948, floated in the nascent Indian financial markets.[36] Just these four private companies (the three managed by Killick and Tata Power) along with the private firm in Calcutta (Calcutta Electric Supply Corporation), India's other major industrial center, still accounted for nearly one-third of all generating capacity in India and most of the industrial consumption in 1960.[37]

Tension between the interests of Maharashtra's privately owned utilities and the board surfaced in later decades as well. One such episode occurred during 1979, as the state government and the board were discussing ways to deal with electricity shortages that had led to episodic power cuts from 1972 onward. In addition to new MSEB thermal and hydroelectric plants at Nashik, Koyna, and Koradi, both Tata Electric Company and BSES had plans to augment their generating capacity. Tata had initially requested in 1965 that the Maharashtra government increase its generating license so it could meet the expanding demand of its industrial consumers, but was assured by the state government that the newly formed public utility would be fully capable of meeting energy requirements in the state. In 1973, the company again requested an expansion of its license to generate electricity. This time, instead of directing this request to the state government, the company approached the Planning Commission and the central government. After first submitting a feasibility report and meeting other technical requirements, the project was cleared by the central government in April 1975 and by the state government in July 1977. The following year, the company started its expansion plans with a loan from the World Bank. In the case of BSES, which hitherto had functioned as a distribution company, its proposed 500 MW thermal power plant represented its first foray in electricity generation. The power plant, to be located in Bassein, would be constructed with German collaboration and World Bank financing. In reviewing the details of these private-sector expansions, a state legislative committee remarked:

> Power is assigned to the Public Sector as a basic policy proposition. In spite of this, Tatas and BSES, both Undertakings in the private sector have been permitted to put up the thermal plans with foreign assistance . . . The State government had proposed that MSEB may be permitted to put up a thermal plant with foreign assistance . . . instead of BSES. This proposal was not accepted by the Centre. Question arises why public sector Undertakings should not be permitted such sanction when the same facility is made available to private sector undertakings.[38]

The report's authors, who are all legislators in the state assembly, imply that the central government acted to protect the interests of influential private companies, India's national capitalists. The committee's report concludes only by asking that the "matter should be reviewed at the appropriate level."

The consequences of keeping the city as a private enclave continued to be felt even during the period of economic liberalization in the 1990s. Because the area of Mumbai and its industrial hinterlands were served by private and municipal distribution and generation companies, neither the industrial consumers that had benefited the most from privatization in Odisha and Delhi nor the urban middle-class consumers that supported privatization because of its promise to provide better service in Delhi were concerned about the fate of Maharashtra's publicly owned electric grid. By the 1990s, the rest of the country was just contemplating a return to the public-private mix that characterized ownership patterns in the colonial period. But Mumbai, along with a small number of other cities around the country (all prominent industrial centers at independence), had never abandoned the private utility model, despite the professed commitment to state ownership that characterized the period.

The story of Bombay and the ability of private companies to influence the nature of India's emerging electricity infrastructure is only half of the narrative in Maharashtra. The other part of the story—about rural electrification—ultimately touches more citizens, and reveals the complex interconnections between local and state politics with infrastructure development. For this, we turn to how the MSEB expanded its reach into rural

Maharashtra and the consequences of this expansion for overall development in the state.

Electrifying Town and Village Maharashtra

Maharashtra's *Third Five-Year Plan*—the first such exercise in the newly created state—established that the goal of planned development was to be the cultivation of an agro-industrial economy. The plan emphasized, on the one hand, increasing agrarian productivity to sustain the nonfarm population, and on the other, raising agricultural and rural incomes so that the countryside could become an expanding market for nonfarm products. The state's Planning Department laid out high agrarian production benchmarks and devoted substantial investments to meeting its twin goals. Of these, the power sector received the largest allocation, 27 percent.[39]

The proportion of funding for the power sector grew further over time. By 1978, 40 percent of the state's planned allocations were devoted to electricity, a sum that many in the MSEB nevertheless considered inadequate to satisfy increasing demand for electricity all over the state.[40] While the heavy emphasis on the power sector was common to many states, what was distinct about Maharashtra's power sector expenditures during the 1960s is that while large sums were given to new hydroelectric projects at Koyna, Purna, Vaitarna, and Sahasrakunda, and new thermal projects farther inland, the state also allocated substantial sums to rural electrification and energizing pump sets. The policy of rural electrification in Maharashtra was not incidental to the overall process of economic planning but the result of strategic decisions taken by the state government. Maharashtra's third plan document makes this point clearly:

> If power generation is to serve the objectives of increased employment, dispersal of industry and building up an agro-industrial society in the rural areas, the temptation of easy disposal of power to bulk users in certain already well-developed areas has to be consciously resisted and for this purpose the transmission lines have to be planned having regard to the above objectives, though at first sight they might appear unduly expensive.[41]

In Maharashtra's vibrant industrial regions in and around Bombay, there was plenty of demand for whatever new electricity either the public utility or the state's private generators could produce. But the planning division cautioned against following this easy route and instead championed the more expensive, time-consuming, and risky model of husbanding new consumers in the state's rural areas. The results of these investments are apparent in the rapid growth of rural electricity consumption in the state. Of the

TABLE 3.1

Category-wise sales of electricity in Maharashtra (million kWh), 1951–1974

Category of consumer	*1951*	*1956*	*1961*	*1966*	*1969*	*1974*	*Increase from 1961 to 1974 (%)*
Domestic	108	166	260	448	603	830	319
Commercial	86	131	198	363	447	562	284
Industrial	816	1,235	1,853	3,297	4,489	6,081	328
Public lighting	7	15	20	39	59	82	410
Traction	271	344	339	414	389	505	149
Agricultural	**3**	**5**	**15**	**90**	**236**	**488**	**3,253**
Public waterworks & sewage	23	31	35	66	114	158	451
Misc.	—	—	—	—	53	106	—
Total	1,314	1,927	2,720	4,717	6,393	8,812	324

Source: Data from Maharashtra Economic Development Council, appendix 8, 60.

seven major usage categories listed in Table 3.1, the gains for agricultural users were the greatest. From 1961 (the first year of MSEB's operations) to 1974, electricity consumption by farmers increased far more than for other categories of consumers. Farmers began the period with the least access to electricity and therefore had the most room to grow. But the growth of electricity use by Maharashtra's farmers was much faster than in other parts of the country. During roughly the same period, agricultural consumption of electricity in all of India grew from 833 kWh in 1961 to 5,006 kWh in 1972, a roughly sixfold increase that is dwarfed by the gains made by farmers in Maharashtra.[42]

We might also note that the dramatic increase in farm consumption does not begin until after the formation of Maharashtra in 1960. From 1951 to 1961, when the Bombay State included both present-day Maharashtra and most of present-day Gujarat, the total annual sales to the region's farmers increased only threefold. These figures can be considered a reliable estimate of consumption and sales patterns since they predate the late-1970s switch from tariffs determined by meter readings to flat-rate tariffs, discussed later in this chapter.

As discussed in the previous chapter, amendments to the national Electricity Act of 1948 weakened the autonomy of the State Electricity Boards (SEBs) and made their functioning more subject to approval by their respective state governments. This enabled politicians to use their state utilities to satisfy political agendas, a tendency that increased gradually from the 1950s through the 1970s. In Maharashtra, one area of interference was in the pattern and rate of rural electrification, which became an important service that politicians could secure for their constituencies. The improved access to irrigation facilitated by the rapid growth of rural electrification was a crucial factor in increasing agricultural productivity, but it also had profound consequences for the functionality of the electricity grid. Without adequate transformer capacity, zealous extensions to the grid caused technical failures that damaged and, in some cases, destroyed electrical equipment, machines, and appliances at the consumer end. In 1969, Sharad Pawar, then a rising member of the Legislative Assembly (MLA) who represented Baramati, a district that was heavily dependent on electricity to run tube wells and lift irrigation pumps, threatened a hunger strike to protest the irregular electricity supplies that damaged pump sets and agro-processing machines. In response, the MSEB representatives testified to the state legislature that the "interruptions and low voltage are largely due to indiscriminate expansion of rural electrification in the past" and further that "the indiscriminate expansion of rural electrification was done mostly on the basis of local pressure."[43] The board argued that since rural electrification yielded political dividends, the vital but less newsworthy labor of augmenting transmission and distribution (T&D) networks was chronically underfunded and underemphasized. During hearings with a legislative committee in 1970, the public utility provided three years of financial statements as evidence of this tendency. Together, these statements demonstrated that in each year, the board requested more for transmission and generation than it received for those tasks. By contrast, the government allocation for rural electrification either met or surpassed the board's requests in each year.[44]

Another consequence of the rapid expansion of the grid throughout the state was chronic electricity shortages, which began in the late 1960s and continued almost annually through the 1970s. Analyzing these episodes of electricity scarcity is useful to gain insight into the nature of distributional conflicts in the sector, as well as their resolutions. They highlight the consistent bias in favor of the expanding ranks of rural electricity consumers. One of the first periods of shortage occurred in 1968, when a poor monsoon reduced the output of the Koyna and Tata hydroelectric stations.[45] Compounding the problem of scarce supplies, construction delays in India's emerging nuclear power industry meant that the quantum of nuclear energy that the MSEB was expecting in 1968 was not available until late 1969. To address the energy shortage, which was expected to last from November 1968 until June 1969 (when the next monsoon would begin), the government initially imposed a 16 percent cut on consumption by all industrial and commercial consumers, and froze the consumption of residential consumers.

In January 1969, the government imposed a 10 percent reduction in residential consumption and disallowed electricity for nonessential purposes, a list that included neon signs, lighting for festivals and marriage celebrations, and state and local government agencies. Notably, the only exemptions were consumers in the defense, radio broadcasting, police, and agricultural sectors.

Electricity shortages continued for much of the 1970s. The causes of the shortages were varied, from delays in the construction of new power plants and poor monsoons that diminished hydroelectric supplies to the poor quality of coal and delays in its supply.[46] On the whole, electricity rationing was most stringent for general industrial consumers, was slightly less onerous for the state's textile mills and other continuous process industries, and only marginal for the category "essential consumers," which was made up mainly of agriculturalists.[47] The shortages and cuts continued through the rest of the decade, reaching a nadir in June 1979, when the state government shut down all general industries and textile mills in the state for five continuous days, a move that a government report acknowledged to be "a drastic measure for which there is no precedent."[48] While neither the state government's energy ministry nor the MSEB was able to supply detailed information about the consequences to the state's economy, media reports suggested that restricting electricity to industries had resulted in daily losses of 20 to 30 crore, in addition to the huge losses to the MSEB in terms of foregone sales.

Subregional Biases in Rural Electrification and Consumption

One striking feature of Maharashtra's impressive rural electrification program is its degree of spatial unevenness. In the first years after the formation of the MSEB in 1960, the utility expended resources laying transmission lines in the districts of Vidarbha, which followed a certain economic logic because the bulk of new thermal generation stations were in the Vidarbha region, and electrifying villages closest to power plants is most cost effective. Marathwada received few electrification resources, both in the early years of utility expansion and later. As politics started to impact the utility's decision making in the late 1960s and 1970s, more of the electrification resources went to the districts of Western Maharashtra, where they became vital for sugarcane farming and cane-crushing operations. So rather than spreading evenly or according to technical or economic logic, the pattern of rural electrification followed, and later reinscribed, deeper political and economic inequalities in the state.

TABLE 3.2

Electrification of irrigation pumps by region in Maharashtra

	Western Maharashtra		*Vidarbha*		*Marathwada*	
	Irrigation pumps energized	*Total connected horsepower*	*Irrigation pumps energized*	*Total connected horsepower*	*Irrigation pumps energized*	*Total connected horsepower*
1960	242	2,844	5,168	15,821	8	43
1965	9,216	45,843	14,587	44,715	1,208	3,846
1970	91,269	459,300	39,659	121,950	28,785	121,100
1975	209,056	1,044,555	92,789	304,815	68,193	299,560

Source: Maharashtra State Electricity Board, *Annual Report*, various years.

The inequality in allocation of electrification resources is manifest in both the rate of village electrification in each region and the number and size of irrigation pumps electrified in each region. Table 3.2 depicts the imbalances in rural electrification across the three primary regions of Western Maharashtra, Vidarbha, and Marathwada during the 1960s and 1970s.

By 1965, the amount of resources going to each region had become more skewed, with the bulk of resources going to Western Maharashtra. The figures showing consumption of electricity by agriculturalists in each region corroborate this tendency. In 1973–74, of all electricity consumed by the farm sector, 74.8 percent went to Western Maharashtra, 14.2 percent to Vidarbha, and only 11.0 percent to Marathwada, which was then and still remains the poorest of the three regions.[49]

In the late 1970s, one decision made by the MSEB had the unintentional consequence of further skewing the distribution of resources in the state among various regions and crops, as well as making any quantitative snapshot of Maharashtra's electricity profile increasingly unreliable. This was the choice to switch from a tariff based on actual consumption of electricity as recorded by a meter to a flat-rate tariff that was based on the size (in horsepower) of the pump. This decision was made in 1977, when Madhav Godbole was chairman of the MSEB, a post he held from 1976 through the end of 1978. In doing this, Maharashtra was following a precedent set by Punjab a decade earlier, as well as recommendations made by the World Bank. Reading the meters of far-flung rural consumers was a time- and labor-intensive activity and there were frequent complaints of malfeasance. In some cases, meter readers accused rural customers of deliberately tampering with meters and, in other instances, utility staff colluded with customers to show a much lower reading of consumption in exchange for payments on the side. The flat-rate system was meant to obviate these myriad problems.[50]

Several decades after his retirement as chairman of the MSEB, Godbole realized that the World Bank's recommendation, which, according to his recollection, were inspired by the electricity policies of much smaller northern European nations like Sweden, were completely inappropriate for India, where distances and populations are much more vast.[51] While the decision may have been made for good reasons, the consequences were far-reaching in further transforming both Maharashtra's cropping and rural electrification patterns. A flat-rate tariff grossly benefited those crops that need intensive irrigation, like sugarcane, by dramatically lowering the cost of electricity per unit. Utility records show that it was farmers in Western Maharashtra who first took advantage of this new facility in pricing, as Table 3.3 demonstrates.

By 1994, almost three-quarters of farmers in Maharashtra paid a flat rate as opposed to a metered tariff for their electricity consumption. In addition

TABLE 3.3

Numbers of high- and low-tension, metered and unmetered consumers by region in Maharashtra

Western Maharashtra

	High-tension (per hp tariff)	*High-tension (metered tariff)*	*Low-tension (per hp tariff)*	*Low-tension (metered tariff)*
1981	505	35	257,419	109,741
1986	657	92	383,635	137,499
1990	820	115	590,807	203,142

Vidarbha

	High-tension (per hp tariff)	*High-tension (metered tariff)*	*Low-tension (per hp tariff)*	*Low-tension (metered tariff)*
1981	23	35	75,873	102,592
1986	23	NA	718,534	157,299
1990	19	57	137,480	172,874

Marathwada

	High-tension (per hp tariff)	*High-tension (metered tariff)*	*Low-tension (per hp tariff)*	*Low-tension (metered tariff)*
1981	6	59	103,277	48,542
1986	9	77	169,231	65,105
1990	43	14	313,616	NA

Source: Maharashtra State Electricity Board, *Annual Report*, various years.

to the regional imbalances in electrification, the benefits of the agricultural subsidy were also imbalanced, this time in favor of larger farmers. A Pune-based civil society organization, Prayas, studied agricultural subsidies and showed that, contra the government claim that subsidies benefit marginal farmers, a very small percentage of mostly larger farmers captures the lion's share of benefits.[52] Only one-fifth of all farmers in Maharashtra either own their own pump sets on private wells or are members of a cooperative lift-irrigation society that requires electricity to lift water from rivers or streams. Of this one-fifth of subsidy beneficiaries, one-quarter of the farmers gain the most.

These two trends together—the spatial bias in rural electrification and the implementation of flat-rate tariffs—help to explain why sugarcane cultivation commands a considerable share of the state's groundwater resources. From 1960–61 to 1997–98, the area under sugarcane cultivation increased from 0.83 percent of the total cropped area in the state, to 2.33 percent of the total cropped area.[53] However, the crop now commands considerably more of the total irrigated area in the state. In 1960–61, sugarcane accounted for 12.33 percent of the state's irrigated land. This figure increased to a high of 18.85 percent in 1994–95 before declining to 13.48 percent in 1997–98.[54] The United Nations Development Program's (UNDP) Human Development Report for Maharashtra notes matter-of-factly: "When irrigation is scarce, Maharashtra has chosen water-intensive sugarcane as its principal cash crop and built an impressive rural socio-political and economic edifice, which is responsible for some one-third of sugar production in the country. More than half of the irrigation goes to sustain a crop on three per cent of [Maharashtra's total] cropped area."[55] We turn next to this sociopolitical nexus in order to understand the political roots of Maharashtra's selective rural developmentalism, a policy regime that is marked by spatial and sectoral unevenness.

Electricity and Agricultural Production, or the Water-Power Nexus

According to the 2001 Indian census, Maharashtra's population is 57.6 percent rural. Some central portions of the state get enough rain for multiple cropping, but most of the state has sufficient rainfall to sustain only one crop. Irrigation, then, has been the key to all gains in agrarian productivity throughout the twentieth century. Agricultural production in Maharashtra has gone through several transformations that have been the result of either technological or institutional changes. In the twentieth century, the first technology to transform the countryside was canal irrigation. The administration of Bombay Presidency constructed large-scale canals in Maharashtra, known as the Deccan canals.[56] Although one purpose of the canals was to provide a measure of insurance against episodic drought and famine, the availability of assured water at the time of an expanding market structure led to further commercialization of agriculture.[57] Most popular among these was sugarcane, which was either made into *gur* (a sugar product produced locally) by the farmers themselves, or processed into sugar by a cane factory. The incentive for those farmers with the necessary resources to switch crops was enormous. Whereas an acre sown with millet or sorghum (the subsistence crops that were the mainstay of this region) would yield Rs. 12 to 13, an acre planted with sugarcane would yield Rs. 618, a fifty-fold increase.[58] From an individual farmer's perspective, then, there were great financial incentives to switch to cash crops for those with the necessary resources. The government, which had financed the expansion of canals, also had an incentive to encourage cash crops, since these would enable the colonial government to recoup its considerable investment. The government therefore created considerable incentives for farmers to lease lands along the canals for sugarcane cultivation.[59]

The first entrepreneurial farmers to cultivate sugarcane were Malis, a community of gardeners who migrated to the region from Saswad near Pune and had the necessary knowledge of cash cropping to take advantage of the opportunities posed by expanded irrigation facilities. Eventually, wealthier members of the dominant agricultural caste, the Marathas, took up sugarcane farming in the decades prior to independence. After independence, Maharashtra embraced a model of cooperation in the agricultural sector. The active involvement of the region's numerically and socially dominant Maratha caste cluster in sugarcane farming and the rise of the cooperative sector together constituted the twin engines driving the transformation of rich Maratha peasants into powerful political actors.

Despite public investments in canal infrastructure, at independence, very little of the total cropped area of the Bombay Presidency was irrigated. The most common form of irrigation was by well, but government canals, private canals, and water tanks or reservoirs were also used.[60] The second significant shift in irrigation patterns and therefore agricultural production came in the late 1960s and 1970s, when rural electrification enabled farmers to efficiently pump groundwater using irrigation pump sets (IPS) and lift water from reservoirs using electrified pumps.

The history of cooperative farming in Maharashtra and the growth of the sugar industry as a mainstay of agriculture and agro-industrialization in Maharashtra have been extensively researched by historians and political scientists. In the next section, I first highlight the ways in which development expenditure on irrigation via electrified tube wells was an important mechanism for these economic and political transformations. Second, I attend to how cooperative institutions facilitated the spread of these new technologies by shouldering some of the risk and cost. Cooperative institutions resolved one of the key dilemmas for bureaucrats charged with allocating scarce development resources. As policy makers and planners in the electricity sector had noted, from a technoeconomic perspective, one of the main obstacles to rural electrification was one of sequencing. Unless a sufficient load, or consumer base, existed in an area, the costs of electrification would not be repaid for many years. So without a political directive and deliberate government subsidy to cover these costs, electric utilities had incentives to concentrate their attention on areas where electricity was already consumed. This was the heart of the problem for rural electrification and one of the early rationales for public ownership of the sector (emphasized in the analysis of Constituent Assembly debates in Chapter 2).

In the rural electrification projects supported by the MSEB and the Rural Electrification Commission, consumers were required to contribute some of the costs of bringing the T&D grid to the field and home. For many, these costs were prohibitive. At the behest of the Rural Electrification Corporation (REC), the Gokhale Institute of Politics and Economics in Pune conducted a cost-benefit analysis of four districts in Maharashtra that had been electrified thanks to REC-funded projects. Based on a large survey of farmers, the analysts reported that even among farmers with sufficient wealth to operate oil engines on private wells, an average of 40 percent reported that an "inadequacy of finance" was a main reason for not shifting over to electricity, even after the distribution lines were brought to the periphery of the village.[61]

Maharashtra's cooperative institutions had several sets of resources that allowed them to lower such costs. As government-supported institutions, cooperatives had the financial means to shoulder some of the costs of bringing electricity to new areas. A 1974 government inquiry commission noted that Maharashtra's sugar factories had collectively spent Rs. 251 million on electrified lift irrigation facilities.[62] While the bulk of these funds undoubtedly benefited cane production for the cooperatives members, there were also likely spillover effects that benefited village households and other farmers. In addition to the financial capital, cooperative leaders staked some of their social capital—prestige within their communities—to convince skeptical farmers of these new technologies.

Cooperative Sugarcane Farming and the Rise of Maharashtra's "Sugar Barons"

Several political, economic, and institutional processes intersect to produce the phenomenon of Maharashtra's "sugar barons" from the period of the 1960s to the 1980s. First is the diffusion of cooperatives across multiple sectors, from credit and housing to irrigation and agro-processing. The impulse to cooperate was deemed a failure in many parts of India, thwarted especially in those places with extreme inequality in land and other asset ownership. In such places, powerful local actors opposed collective-ownership arrangements. In Maharashtra, however, the cooperative movement was enormously successful, supported both from above by influential policy makers and politicians and from below by local landowners who saw the potential of cooperative institutions as conduits for credit.

Cooperative processing institutions became the vehicle for the expansion of the state's emerging sugar industry. During the colonial period, the dominant sugarcane-growing regions of India were located in North India, in present-day Bihar and Uttar Pradesh, where the subtropical climate was naturally suited to cane cultivation. In the north, the factories that crushed the cane into sugar were almost entirely privately owned. Although there were some British-owned companies, by the early twentieth century, the bulk of North Indian sugar factories were owned by Indian firms, including the prominent family business houses of Shriram, Birla, Walchand Hirachand, and Thapar.[63] The sugar factories resolved the central coordination problem of sugar production—which requires getting cane to the factory at precisely the time when its sugar content is highest—not through a plantation economy of the kind that emerged in many other colonies, but through a system of intermediaries. "Landlords, money-lenders, and rich peasants . . . derived their strength not so much from their connection with the factories, or their entrepreneurial skills in exploiting new market opportunities as from their traditional position of political and economic domination in rural society."[64] Dense, hierarchical social relations between smaller farmers and those who came to own and operate the sugar mills facilitated the process of transferring cane to the mills at just the right moment.

It was only as irrigation facilities spread throughout western India that sugarcane farming became viable and lucrative there, in what are India's tropical zones. Given the uneven supply of canal waters, even before widespread electrification brought down the cost of well irrigation, farmers who grew sugarcane on canal tracts relied heavily on wells to ensure adequate irrigation.[65] Electrification was doubly important for sugar farming, which requires large amounts of water delivered at a precise time in the production cycle. In the semiarid Deccan, electric pumps to lift surface water and groundwater became an essential catalyst for Maharashtra's burgeoning sugarcane and sugar industries.

In Maharashtra, those who cultivated cane were themselves entrepreneurial and capital-rich farmers—first from the Mali caste and later the Maratha caste—who could take the risk of leasing land for a premium along the new canals. In the 1930s, cane-growing Malis switched from producing gur to processing their cane into refined sugar.[66] Sugar had many advantages over gur, most importantly its longer shelf life and higher prices. Whereas gur could be produced in a decentralized fashion by individual families of farmers, the costs of refined sugar production required the establishment of larger factory enterprises. Rather than selling their cane to the existing private crushing companies in the region, which had built a series of small factory-run plantations, the Malis in one region solved the problem of coordinating cane production for factory processing by starting their own crushing cooperative in 1934. The cooperative was extremely successful, expanding operations on several occasions and repaying loans ahead of schedule.[67]

Thanks to the success of the Mali factory, other farming communities also chose to bypass the private factories and establish their own factories along cooperative principles. One of the earliest was set up in 1948 by cane-growing Maratha agriculturalists with the organizational and conceptual help of D. R. Gadgil, who is sometimes called the father of the cooperative movement in Maharashtra.[68] V. Vikhe, the main leader of the cooperative, organized his fellow farmers to contribute half of the share capital, convinced the Bombay State Cooperative Bank to raise the other half of the share capital, and arranged a loan from the Industrial Finance Corporation to finance the costs of equipment and construction.[69]

The cooperative sector received a major boost in 1954, when the provincial government of Bombay convinced the central Indian government to reserve all future licenses for sugar factories for the cooperative sector. In itself, this could be read as testament to the growing clout of the commercialized peasantry, most of whom came from the upper echelons of the Maratha community. Cautioning against the danger of reading history backward, however, one scholar of the sugar cooperatives suggests that the early cooperatives were the product of experimentation; it was only over time that they came to represent a "major power base" of the rural Maratha elite.[70] The Bombay-based industrial houses were vocal but ultimately ineffectual opponents of the new policy. Cooperative factories allowed landowning and commercial farming communities to challenge the political and economic strength of urban-based industrial capital. Thanks to the resources of the state banking and credit societies, these landowners also were able to overcome the constraints of capital scarcity. As one scholar notes, the "sugar cooperative, therefore, became an important vehicle for this transformation of the Maratha cane-growing community from that of a subordinate peasant community to becoming the dominant agrarian class."[71]

The existence and pattern of cooperatives in Maharashtra were integral to the Congress Party's continued dominance in the state, even as effective national and local opposition parties in the country challenged the hegemony of the Congress. The first state to elect a non-Congress government was Kerala in 1957, which elected a Communist government; in subsequent decades, elections all over the country became multiparty contests in which the Congress frequently lost. Maharashtra, however, remained a Congress stronghold. The brief period of non-Congress rule in Maharashtra in the aftermath of Indira Gandhi's "Emergency" in the late 1970s seems only to confirm the strength and influence of Congress politicians in the state's politics. When the Janata Party won the national elections in 1977, Maharashtra was one of five states in which the Congress retained a large measure of its former power.[72] The Congress also was defeated in most of the state elections that coincided with or followed the national elections. In Maharashtra, however, the only way that a non-Congress government could come to power was after a local Congress politician, Sharad Pawar, defected from the party along with a large number of other ex-congressmen. The Congress itself has split several times into factions organized around powerful leaders, but barring the brief period in the late 1970s, these splits did not allow opposition parties to come to power in the state until the 1990s.

Cooperative institutions were critical to Congress success because they provided avenues for the party to build durable alliances at the local level. Cooperatives were also the means by which the state government could garner support from influential segments of the countryside for agro-industrial development.[73] There were multiple channels linking the Congress Party to the cooperative institutions. Congress politicians were among the first to establish new agricultural cooperatives, relying on state credit to do so. Furthermore, the Congress Party inducted new political elites from among the cooperatives.[74] Rural political aspirants who first rose to local prominence as directors, managers, and board members of cooperative institutions were drawn into the Congress fold, bringing with them hard-won local political experience and ready-made constituencies.[75] An analysis of the Maharashtra legislative assembly in office from 1967 to 1972 determined that out of the 263 legislators who responded to the questionnaire, 159 (or 60.7 percent) of them were chairmen, directors, or secretaries of credit, sugar, or other state cooperative institutions.[76] Already sizable in 1960, the contingent of cultivators and landlords within Maharashtra's legislative politics only grew more substantial over time. In 1960, one-third of legislators identified as farmers; by the fifth legislative assembly in 1978–80, the percentage that identified as such had grown to almost one-half. By contrast, the presence of traders and industrialists fell from 20 percent in 1960–62 to just under 7 percent by 1978–80.[77] Even more striking was farmers' dominance in ministerial positions. The percentage of cabinet ministers who identified as agriculturalists increased from 51 percent in 1960 to 55 percent in 1978. The second-largest category after this was lawyers, who occupied 20 percent of ministerial posts in 1960 and only 10 percent in 1978.[78]

There were other vectors by which farmers' interests came to be represented in state politics. In the 1970s and 1980s, a significant social movement of farmers grew increasingly effective.[79] Throughout the 1980s, farmers organized to demand improved terms of trade that included subsidies for inputs like fertilizers, electricity, and seeds, and higher prices for farm output. There were several epicenters of the movement, including in Punjab and Haryana, Andhra Pradesh, and Maharashtra. Over time, the various regional incarnations of farmers' lobbies were united under a single umbrella organization.

In Maharashtra, the most influential farmers' organization was the Shetkari Sanghatana, led by the charismatic Sharad Joshi, who came to farming relatively late in life after a diplomatic career that included stints at the United Nations in Europe. The farmers' most effective means of protest included demonstrations that involved hundreds of thousands of peasants and blockades of railroads and entire villages to government officials and politicians; in Maharashtra, farmers also withheld crops like onion, cotton, and tobacco from the markets, causing a spike in prices. Movement leaders also urged farmers to stop paying taxes, electricity bills, and bank loans.[80]

As suggested previously, the state in Maharashtra had deep roots in the countryside. One consequence of this is that throughout the 1960s and

1970s, Maharashtra devoted more electrification resources to rural development, particularly in the western cane country, than most other state governments. This resulted in a very different pattern of rural electrification—and, more important, the electrification of irrigation pumps—than that of other states. To judge how important electricity had become to agricultural production, we can compare state-level data on electrified pumps.

The figures given in Table 3.4 for potential electrified pumps were calculated by the Ministry of Power on the basis of groundwater stores, area under cultivation, and the extent of alternative irrigation supplies, such as from canals. In some states (among them Odisha, Bihar, and Uttar Pradesh), less than a third of the potential was met. Maharashtra is one of two states (the other is Karnataka to the south) where nearly all of the potential for electrifying irrigation pumps was achieved by 1993.

Beginning in the late 1970s and continuing through the 1980s, the increases in rural electricity consumption were attended by reductions in the rates farmers paid for these services. In 1978, Sharad Pawar, chief minister at the time, announced a reduction in the electricity rates for farmers from 29 paise per unit to 20 paise per unit, and a corresponding reduction in the

TABLE 3.4

Village and irrigation pump electrification in India as of 1993

State	*Total villages (1981 census)*	*Number electrified as of March 1993 (% of villages)*	*Estimated potential number of electrified pumps*	*Number electrified as of March 1993 (% of potential)*
Andhra Pradesh	27,379	27,358 (99.9)	1,600,000	1,398,049 (87.4)
Assam	21,995	21,481 (97.7)	200,000	3,675 (1.8)
Bihar	67,546	47,498 (70.3)	1,000,000	261,100 (26.1)
Gujarat	18,114	17,892 (98.8)	700,000	512,780 (73.3)
Haryana	6,745	6,745 (100.0)	430,000	396,639 (92.2)
Himachal Pradesh	16,807	16,761 (99.7)	10,000	3,755 (37.6)
Karnataka	27,028	26,483 (98.0)	850,000	869,461 (102.3)
Kerala	1,219	1,219 (100.0)	300,000	265,224 (88.4)
Madhya Pradesh	71,352	65,468 (91.8)	1,300,000	1,003,900 (77.2)
Maharashtra	39,354	39,106 (99.4)	1,800,000	1,760,976 (97.8)
Odisha	46,553	32,682 (70.2)	500,000	61,428 (12.3)
Punjab	12,342	12,432 (100.7)	700,000	639,343 (91.3)
Rajasthan	34,968	28,460 (81.4)	600,000	439,120 (73.2)
Tamil Nadu	15,831	15,822 (99.9)	1,500,000	1,402,858 (93.5)
Uttar Pradesh	112,566	84,256 (74.9)	2,400,000	694,902 (29.0)
West Bengal	38,024	28,455 (74.8)	500,000	94,710 (18.9)

Source: Ministry of Power, *Annual Report 1994/95* (New Delhi: Government of India, 1995), 24–25.

rate for nonmetered electricity consumers from Rs. 180 to Rs. 125 per hp.[81] At the same time, the government also removed the upper threshold for nonmetered supply with the explicit aim of facilitating cooperative irrigation societies. The new policy required the government to subsidize the utility by an additional 12 crores each year. The costs of electricity for domestic, commercial, and industrial consumers rose gradually over the decade, but costs to farmers were kept constant. The result was a system of cross-subsidies; nonagricultural electricity consumers (as well as consumers in Bombay, whose prices were always higher than in other parts of the state) offset the costs of electricity used by those farmers who could afford to run their own pump sets or who benefited from cooperative lift irrigation facilities.

As previously indicated, there were technical and financial consequences to the utility from these large extensions to the electricity grid, particularly when investments in the T&D systems were inadequate. Yet despite the rapid increases in rural electrification and the reduction in the tariffs paid by farmers, the condition of the utility improved in the 1980s thanks to expansions in the state's generating capacity. The MSEB added a new gas-fired power plant at Uran, which started producing electricity in the 1980s, and a new coal fired installation at Chandrapur in eastern Maharashtra, which started producing electricity in 1983 and continued adding capacity over the next several years. The capacity additions allowed the board to sell power to neighboring states, including Gujarat, Goa, and Andhra Pradesh. These sales provided another way for the board to make up for the unprofitability of rural electricity provision. The phenomenon of captive power generation, in which private companies left the utility in favor of generating their own electricity supplies, was also not as extensive in Maharashtra as in other states, which meant that the MSEB retained more of the state's high-paying industrial consumers than other public utilities were able to manage.[82] At the end of the 1980s, the MSEB was the only electricity board in the country that made a commercial profit.[83]

Maharashtra During Market Reforms

In the early 1990s, Maharashtra was considered by some policy makers and energy experts to be an ideal candidate for utility privatization. First, one of the main justifications for continued state ownership—making access to electricity more inclusive by taking up the expensive task of rural electrification—was at least notionally no longer valid since the state claimed to have achieved full electrification by the late 1980s. Second, the utility was among the most profitable of all electricity boards in India, and though T&D losses were rising, the MSEB's technical efficiency was still better than in other utilities, so finding buyers was expected to be relatively easy. With limited capital resources, unbundling the board and strategically selling off pieces of it to private companies was a way for the state to finance capacity additions. All of these facts coincided with a growing reluctance on the part of international financial institutions, which had backed most of India's earlier generation expansion, to continue to fund state-owned companies that often operated with deficits and growing debt.

Maharashtra's political and policy establishment were also considered reformist because of the way the state had embraced other aspects of market reforms. Within the energy sector, as early as the 1980s, the government had indicated a willingness to embrace new models of growth. For much of the early 1990s, P. Abraham served as chairman of the MSEB. Like Jagdish Sagar, the chairman of Delhi's utility when it privatized its distribution network a decade later, Abraham was an adroit career bureaucrat who embraced the new climate of liberalizing India.[84] He served in various capacities in the sector, as energy secretary to the Maharashtra government from 1986 to 1989, as chairman of the MSEB from 1989 to 1991, and as power secretary to the government of India from 1994 to 1997. In 1989, the government of Maharashtra started lobbying the central government to approve new gas-based power projects. These would have required supplies from gas and oil terminals at a time when the demand for natural gas and oil was already exceeding its availability in India. Although the central government had not yet amended the electricity act to permit private investments in the sector, which would not happen for another two years, Abraham told the press that the state was "nonetheless preparing grounds for such eventuality."[85]

For all of these reasons—the government's ostensible support for market reforms, the financial health of the MSEB, and the fact that electrification was nearly complete—the MSEB was considered to be a strong candidate for privatization. One proposal that emerged in the ongoing discussions between the SEB and the government was to restructure the utility to allow private equity participation, to float the SEB via the Bombay stock exchange. Whether through this method or direct sale of assets to a single investor, the energy policy community determined that "it may be that a real stab at privatisation is an idea whose time has come. If so . . . then the MSEB is as good a place as any to start."[86]

Within the government, though, calls for distribution privatization were more muted. I argue that one important explanation for this is that privatization threatened to dismantle the way that subsidies, electricity supplies, and electricity cuts were being allocated within the state. By the end of the 1990s, power subsidies for Maharashtra's agricultural consumers were among the highest in the country.[87] So while many in multilateral institutions and policy circles, particularly in New Delhi, had begun to champion privatized utilities, in Maharashtra, there was a continuing belief that the state had an important role to play. As Maharashtra's energy minister in the 1990s, Dilip Walse-Patil, put it, "in a democratic welfare state, you can't think about everything in terms of cost and profit."[88] Instead, electricity pricing and allocation should follow social, developmental, and arguably political logics rather than purely technocratic or corporate ones.

As I suggested earlier, the government's opposition to privatizing distribution did not constitute an ideological opposition to market reforms. In fact, Maharashtra's policy record in the 1990s shows a government eager to innovate in other areas of energy-sector reforms. In the main, these were reforms that promised increases to the state's generating capacity without threatening the existing distribution matrix and price structure. The first was the entry of private firms to build new gas- and coal-based power plants. Initially the state government invited private-sector bids for two new projects—a gas plant at Uran and a coal plant at Khaperkheda, near Nagpur—to which fourteen private firms responded.[89] At the same time, the government of Maharashtra also identified seven other sites for new generating plants.

Of all the new private investors in the sector, though, the most famous became the consortium led by the US-based Enron Corporation. Almost from the start, details of the investment project were obscured in controversy, and imbricated in one political storm after another. As with Enron's spectacular collapse in the United States, in India, too, journalists, academics, and outraged citizens aired their perspectives and positions in a wide range of books and articles, which collectively point to but never fully illuminate the accusations of political corruption, bribery, and financial malfeasance that circulate around Enron's investment in Maharashtra.[90] These investigations and reportage, on which the following narrative is based, agree on the timeline of events, but not on how to assign blame for the project's many failures.

The MSEB had initially identified the site at Dabhol in Ratnagiri District for a thermal power plant in 1989, a proposal that the Central Electricity Authority (CEA) declined to pursue in the public sector.[91] At the time, New Delhi and multilateral funding bodies were rethinking how to increase capacity in the energy sector, and were increasingly encouraging private firms rather than public investments. In 1992, a joint delegation from Enron and General Electric visited first New Delhi and then Bombay, and after three days in the state capital, signed a Memorandum of Understanding (MOU) with the MSEB for a 2,000 MW thermal plant that would use imported liquefied natural gas (LNG) as the fuel source. In 1990–92, the generating capacity in the entirety of the state, including the MSEB, central power utilities, and private firms, was only 8,729 MW, so the MOU represented close to a 25 percent increase in overall electric capacity.[92] At the time, Maharashtra was projected to have "peak" energy deficiencies, that is, shortages during those times of the day and days of the year of greatest simultaneous electricity demand. But for the most part, the public and private utilities were able to meet the base demand for electricity, barring those times of equipment failure or system destabilization. The MOU signed between the MSEB and its foreign partners, however, committed the utility to pay for the total output of the power plant regardless of whether it was consumed or needed, in essence, adding to the state's base capacity and at a rate that was higher than the cost of supply from the MSEB's existing hydro- and thermal power plants. These two facts, together with the use of costly imported fuel rather than local coal, the considerations of foreign exchange commitments, and the absence of an open bidding process, were targets for criticism from the press, public intellectuals, and opposing political parties.[93]

After the initial MOU was signed at the state level, the CEA in New Delhi granted its clearance of the project in late 1993. Earlier that year, the Indian Ministry of Finance approached the World Bank for project funding, but the Bank's negative assessment of the project, which was leaked to the press in 1993, noted that although the Bank favored the "government's private power initiatives," its economic assessment was that the "project is not viable."[94] Nevertheless, the MSEB and the government of Maharashtra pushed ahead. By early 1994, the issue became politically contentious. While campaigning for the 1995 state assembly elections, politicians from the regional Shiv Sena and the national Bharatiya Janata Party attacked the Dabhol project on economic nationalist grounds and promised to cancel the agreement if elected.[95] In particular, these two parties, as well as the Swadeshi Jagran Manch (a social movement affiliated with the Hindu nationalist Rashtriya Swayamsevak Sangh that emphasizes economic issues), criticized the guarantee signed by the government of Maharashtra and the counterguarantee supplied by the government of India to secure the obligations of the MSEB in case of late payments or default.

As promised, after winning the elections, the new coalition government formed a committee to reevaluate what had become Maharashtra's largest foreign investment to date. Within just eleven days, the committee submitted a report that recommended proceeding with the project under slightly altered terms.[96] On the one hand, the renegotiated deal lowered the capital costs of the project and consequently the financial obligations of the board to the Dabhol Power Company (DPC), but on the other hand, the project size was increased by nearly 200 MW, which resulted in a net financial loss for the MSEB, particularly if the extra power had to be paid for but for which there was no demand.[97] Even as construction for the project commenced in late 1995, the remaining political parties in the state, as well as local communities near the proposed location of the plant in Ratnagiri District continued to protest.[98] A 1999 Human Rights Watch report amassed evidence that state and local authorities had responded to these protests with undue levels of violence.[99]

The first part of the two-phase power plant was a 740 MW naphtha-based plant that was commissioned and started generating electricity in 1999. The second part was a much larger 1,444 MW, LNG-based plant that was in the midst of construction in 2000 and 2001 when three events interfered. First, after making full payments for Dabhol's electricity from June through November 1999, the MSEB started making partial payments from December 1999 to May 2000, and then stopped paying the DPC entirely from June 2000 onward. The DPC invoked the state guarantees to collect payments directly from the government of Maharashtra's reserve fund. Second, elections in 1999 returned the Congress–Nationalist Congress Party alliance to the state government. The new government appointed a committee to again review the Enron project and recommend a way out of the impasse that all parties were in as of 2001. Known after its chairman as the Godbole Committee, the report unearthed multiple problematic decisions taken by the government. The introduction mentions the "numerous infirmities in the process of approvals, which bring into question the propriety of the decisions," a statement that can only be an oblique reference to the lack of transparency that characterized the process from the initial bidding to its subsequent renegotiation.[100] The report recommended a complete restructuring of the agreement between the DPC and the MSEB, including limiting dollar-denominated liabilities, lowering the interest rates on outstanding debt, reducing the tariff to be paid by the MSEB, and finding buyers for DPC power outside of the state.[101]

A final event that had direct bearing on the DPC's fortunes was Enron's financial collapse in the United States, ending in the company's bankruptcy filing in December 2001. Enron ultimately sold its $3 billion equity stake in Dabhol for $20.4 million in 2004.[102] The two other foreign owners, General Electric and Bechtel, finally settled their claims on Dabhol in 2005, when both companies sold their stakes to a subsidiary of the MSEB, the Maharashtra Power Development Corporation. The ironic aftermath of Enron's investment, which was intended to be one of the first and largest private-sector power projects in postliberalization India (and also one of the largest gas-fired power plants in the world), was the reconstitution of the Dabhol plant firmly in the public sector. The newly christened Ratnagiri Gas and Power Company was formed in 2005 to take over the assets of Dabhol, and is jointly held by the National Thermal Power Corporation (NTPC), Gas India Limited, the MSEB, and a handful of public-sector energy and financial institutions.

The government's second policy to increase generating capacity was the promotion of biofuels—in particular, the waste materials from sugarcane production, or bagasse. In addition to enhancing energy production, in Maharashtra, promoting biofuels also became useful for shoring up what by the 1990s had become a faltering sugar industry. The decline in productivity and profitability of the sugar cooperatives was pinned to aging equipment and competition from new private sugar factories in Tamil Nadu that had more efficient production technologies. In addition, Maharashtra's sugar sector suffered the consequences of its own rapid success in the sense that the government authorized too many new factories in close proximity to existing ones, forcing the cooperatives to compete for sugarcane supplies. Insufficient sugarcane meant that machines sat idle and factories were operated at far less than total capacity.

Early discussions of biofuels in India occurred in the early 1970s, initially spurred by domestic energy scarcities and later by global price increases in the wake of the 1973 oil embargo.[103] Circulating around these discussions of alternative fuel sources were models for decentralized energy systems, very different from the large, interlinked model of the energy grid that had guided Indian energy policy since the 1948 Electricity Act. Biofuels held out the promise that numerous small power plants could serve discreet, dispersed locales and contribute any excess energy to the larger grid.

These early discussions and government enthusiasm for alternative energy production and distribution impacted the sector only at the margins. Throughout the subsequent decades—and into the present—the main emphasis in Indian energy policy continued to be large thermal plants integrated in a comprehensive grid. In the early 1990s, though, there was an additional stimulus to biofuels when the World Bank and other large institutional investors began focusing on alternative energy as a new model for environmentally sustainable economic development. These trends were institutionalized in 1992, when the UNDP, World Bank, and bilateral aid agencies from the Netherlands and the United States formed the Asia Sustainable and Alternative Energy Program. The initial goal was to increase the share of alternative energy funding to 10 percent of all World Bank lending to the Asian power sector.[104]

In 1993, the Ministry of New and Renewable Energy (MNRE) publicized a new initiative to regulate and encourage bagasse-based energy production.[105] To demonstrate to mill owners the benefits of channeling the waste materials from sugar production to generate electricity, the Indian government with financial support from the United States Agency for International Development (USAID) launched a demonstration program that financed bagasse plants in nine sugar mills around the country. The program lasted for five years, from 1997 to 2002, during which time the production of electricity from sugar waste materials grew severalfold. Technically, bagasse-based electricity generation is an example of cogeneration, in which steam and electricity are generated from the waste materials (crushed sugarcane) left over from production—in this case, sugar production. By updating to new boilers and turbines, larger sugar mills could produce enough energy to run the sugar mill itself, as well as an excess that could be sold back to the grid. Advocates of bagasse cogeneration note that India as a whole, which is second only to Brazil in sugar production, could add an additional 5,000 to 5,600 MW of energy to its current capacity of 145,600 MW.[106]

USAID's demonstration projects in nine Indian sugar mills succeeded in widely disseminating technical and financial information about bagasse cogeneration, as well as showing that firms that made the switch reaped steady revenues from electricity, particularly attractive amid sugar's commodity price fluctuations.[107] In Maharashtra, the government embraced the bagasse program because it promised new revenue streams for politically important cooperative institutions. But the very fact that sugar mills were cooperatively managed became a serious obstacle to the emergence of bagasse cogeneration. Since, in a cooperative, profits from each year's sugar sales are redistributed to the members of the cooperatives as higher prices for the next yield of sugarcane, cooperative mills have less capital than private mills to invest in new capital acquisitions. Cooperatives are also less able than the private sugar mills to raise funds from private banks for these substantial capital costs.

To offset these disadvantages, the government in Maharashtra provided a series of incentives to encourage bagasse cogeneration. The Maharashtra Electricity Regulatory Commission (MERC) was the first in the country to issue a tariff policy for bagasse cogeneration.[108] According to the petition filed by one public advocacy group, Pune-based Prayas, the tariff rate that MERC set for bagasse cogenerators—which was the tariff at which the MSEB would purchase electricity to distribute through its grid—was excessively generous toward the sugar mill owners.[109] Prayas noted that through its generous tariffs for bagasse producers, MERC was forcing Maharashtra's electricity consumers to subsidize the mills. These various incentives in financing and remunerative tariffs facilitated the rapid expansion of the bagasse sector. According to the Maharashtra Energy Development Agency, which is in charge of promoting renewable energy projects, in 2011, thirty-three cooperative sugar factories had a collective installed capacity of 480 MW, and another ten private factories (which tend to be larger in size) had an additional installed capacity of 191.5 MW.[110]

During the 1990s, the state in Maharashtra was willing to experiment with its electricity sector in an attempt to find new sources of generating capacity, as the discussion of Enron and bagasse show. Eliminating agricultural subsidies, however, was not seriously considered by any of the state's political parties, even though these were being touted as a main problem for state finances. In terms of the overall structure of Maharashtra's electricity sector, the central government, which all along had advocated privatization and restructuring, finally forced change in reluctant states like Maharashtra. The central mandate to unbundle the integrated public utilities into separate generation, transmission, and distribution companies came into effect in 2003. Two years later, in 2005, the MSEB was formally separated into four successor firms: a single holding company and separate companies for each of the three functional areas. Several years prior to these structural changes, the government set up the Maharashtra Electricity Regulatory Commission in 1999. As I discuss in Chapter 2, the quality of the regulatory commissions across India varies widely, as does the quality of regulation in other sectors. Maharashtra's is one of the better-off regulatory commissions in terms of the transparency of its operations and the extent to which it is responsive to the broader public and not merely the private companies and utilities. During the early years of its operations, however, MERC too was unable to recalibrate tariffs to reflect the cost of supply; farm subsidies endured.

Conclusion

This chapter explores the distinct pattern of Maharashtra's electrification, and further reveals the connections between earlier histories of electrification and the policies that the state government pursued during the 1990s period of economic liberalization. From the 1960s to the 1980s, there was a steady flow of resources and preferential treatment of rural electricity consumers located in and around the emerging sugarcane tracts in Western Maharashtra. Given these political entanglements with electrification, when the Indian central government began to advocate privatizing the public electricity system in the early 1990s, the government in Maharashtra chose instead to pursue energy-sector reforms that promised to support rather than dismantle the existing matrix.

In all of this, though, the pattern of electrification has not been universal. Resources for rural electrification have been concentrated among particular regions of the state (Western Maharashtra) and rural citizenry (medium and larger landowners, and those with connections to rural cooperatives). Electricity in general, and rural access to electricity in particular, is still not conceived of as a social good but rather as a tool to further a particular pattern of economic development. The consequences of this continue to be seen in the uneven patterns of access to electricity across Maharashtra. Many parts of the state—including urban centers like Mumbai, Thane, and Pune, but also rural districts like Satara—have high rates of household electricity access. There remain pockets of the state, like Gadchiroli in the far eastern portion of the state and Nandurbar in the far north, in which access is still limited to fewer than 60 percent of the population.

4

Extractive Industrialization and Limited Electrification in Odisha

Throughout the early 1990s, an aggressive global consensus advocated that capital-scarce economies with growing energy needs should turn to private sources to set up new generating plants. Within a few years, however, it had become clear to many of these same advocates that the electricity sector could not be fixed merely by allowing private companies to sell power from new plants to existing, state-owned electricity companies. Since that realization, the central Indian government and multi- and bilateral donor organizations began advocating privatization of electricity distribution utilities. Despite this strong urging and the popularity of utility privatization in many other parts of the world, most Indian state governments resisted privatization. By 2004, almost a decade after the first national-level conference was organized to promote distribution privatization in Jaipur, Rajasthan, only two regional governments, in Odisha and Delhi, had privatized distribution systems. In 1993, the government of the eastern state of Odisha, then and still today one of India's poorest states, started a process that five years later would result in the complete overhaul of the state's electricity system, the first such privatization of any utility in India. This chapter analyzes the historical context and political-economic conditions in which privatization unfolded in Odisha.

For several reasons Odisha was an unlikely candidate for such a radical experiment. Not only was the state among the poorest in India by many indicators of human, social, and economic development, but rural electrification had proceeded much less rapidly in Odisha than in other parts of India. As we see in Chapter 2, rural electrification was among the foremost concerns driving nationalization of the sector in the late 1940s, prompted by the track record of colonial-era private utilities that had largely bypassed Indian villages. If the completion of publicly subsidized rural electrification was the marker of a mature electricity system, a state with a more advanced electric grid—like Punjab, Haryana, or Maharashtra—would have made a likelier candidate for privatization than Odisha, which still had large numbers of unelectrified villages. And although the finances of the state's public utility were strained by debt and insufficient revenue, by neither technical nor financial measures was the Odisha State Electricity Board (OSEB) the worst-off of the State Electricity Boards (SEBs).[1]

Why had electrification in Odisha proceeded more slowly than in most other parts of India in the decades after independence? And why, despite the incomplete project of rural electrification, did the government of Odisha privatize its utilities in the 1990s, returning the state to the same condition that the nationalizers had sought to subvert by creating publicly owned utilities in the 1940s and 1950s? In this chapter, I propose a set of related explanations to these two questions. In doing so, I argue that the state government's seemingly surprising utility privatization was actually a reasonable extension of the logic of development that had guided the state from the time of its formation to the present.

Most accounts for Odisha's unlikely privatization focus on the lending terms imposed by the World Bank. Studies that do mention the importance of historical features of Odisha's political economy do so only in passing. As I discuss in the introduction, such a view parallels the larger tendency in the literatures on market reforms and economic liberalization to overlook the way that historical patterns can discipline and structure contemporary possibilities. While the role of actors like the World Bank, the Asian Development Bank, and the Indian central government in shaping economic liberalization at the state level should not be discounted, the analysis of this chapter focuses on the provincial political economy as it took shape in the decades after independence. My research, which includes interviews with some who were closely involved with the privatization, suggests that key leaders in the state were already receptive to the idea of privatization, and that the World Bank's lending conditionality served to strengthen a resolve already in place.[2] Privatization of the electricity sector—with all of the promises that it afforded—was supported by a political elite who perceived it to be a strategy that could advance an agenda of large-scale industrialization.

Utility privatization was possible in Odisha because the previous history of infrastructure development in the state had focused almost entirely on industrial consumers. The limited nature of Odisha's rural developmental efforts from the 1950s to the 1980s provided the enabling conditions for its exuberance as a market reformer from the early 1990s onward. Whereas rural electrification had become a primary means to improve agricultural productivity in many Indian states, particularly during the Green Revolution and after, in Odisha, electricity consumption by farmers remained low throughout the postindependence period and into the present. Many parts of India that have larger indigenous populations also have fewer public services, not just electricity but also roads, telecommunications, and primary health and education facilities.[3] But in Odisha, it is not only districts with high concentrations of indigenous groups that have low rates of electrification; household electricity access is lower than the national average even in districts with relatively small indigenous communities. And importantly, the farm sector throughout the state consumes a much smaller proportion of electricity that in other parts of India. Therefore, farmers' groups who were among the first to resist privatization in other parts of India were silent in Odisha. The lack of rural developmentalism in Odisha's earlier decades, then, created the possibility for successful privatization once the Indian national government opened the sector to private capital and private ownership in the early 1990s.

Although Odisha's population is primarily rural, agriculture has not been the focus of the state's development strategy. Perhaps one explanation for this is that no significant agrarian political movements of the kind seen elsewhere in India have successfully translated the numerical majority of the countryside into political power in the state legislature. In postindependence Odisha, the apex of political authority was concentrated within a primarily urban political class. For these political elite, the natural resource wealth of Odisha's upland regions held out the promise of rapid growth through extractive industrialization. Odisha's politicians from the late colonial period onward had expected the state's mineral wealth to furnish the basis for industrial development, but this hope was constrained by the scarcity of capital resources. Once the central government opened up vital sectors of the economy to private capital beginning in the early 1990s, Odisha's industrialization program was reimagined and pursued with new vigor. In all of this, electricity privatization was useful because it signaled the state's commitment to market reforms and it promised to remove the obstacle of insufficient power capacity.

Although most of the state's population is nonurban, a high degree of social fragmentation along caste and tribal lines has made it difficult for a unified rural movement to emerge. In Maharashtra, for example, a consolidated rural coalition built on the dominant Maratha caste steered state resources toward the countryside from within the state apparatus. The rural political coalition both put pressure on the state's public utility to expand rural electrification and channeled financing to pay for pumps and lift irrigation through the burgeoning cooperative sector. Chapter 5 on Andhra Pradesh presents another mode by which the rural areas influenced development expenditure. There, farmers' groups mobilized outside of the state to lower the rates of rural electricity. Odisha is one of the few states remaining in India without a peasant caste mobilization of the kind that successfully diverted development resources to the rural economy in earlier decades, and that, in more recent decades, resist market reforms when they encroach on rural interests. These actors were absent in Odisha.[4]

In Odisha, then, electricity sector privatization was the product of a confluence of factors: a lack of political opposition to privatization; a motivating belief in extractive industrialization among the state's political elite; and, finally, the ascendancy of a new global model that promoted private-sector-led development, particularly in energy production. Explaining privatization as the product of external pressures alone (from either the World Bank or the central government) is insufficient. State-level political economic variables allowed such external pressures to find their mark. The remainder of this chapter presents an analytic history of Odisha's political economy conjoined to a history of electrification in the state.

The Regional and Social Bases of the State in Odisha

Like most other Indian states, politically modern Odisha is a composite of regions with differing political histories. Odisha's coastal districts, those in and around the capital city of Bhubaneshwar, were governed as parts of several different British presidencies and were consolidated into an independent colonial unit under the name of "Orissa" only in 1936. The inland hill districts, by contrast, were composed of princely states with a range of tutelary relationships to the British crown. Following the formation of Orissa after independence (the state was renamed "Odisha" in 2011; the remainder of this chapter uses the new spelling), the sharpest line of differentiation continued to be between the former colonial districts of the coastal lowlands and the highland districts of the interior. At the point of their political union

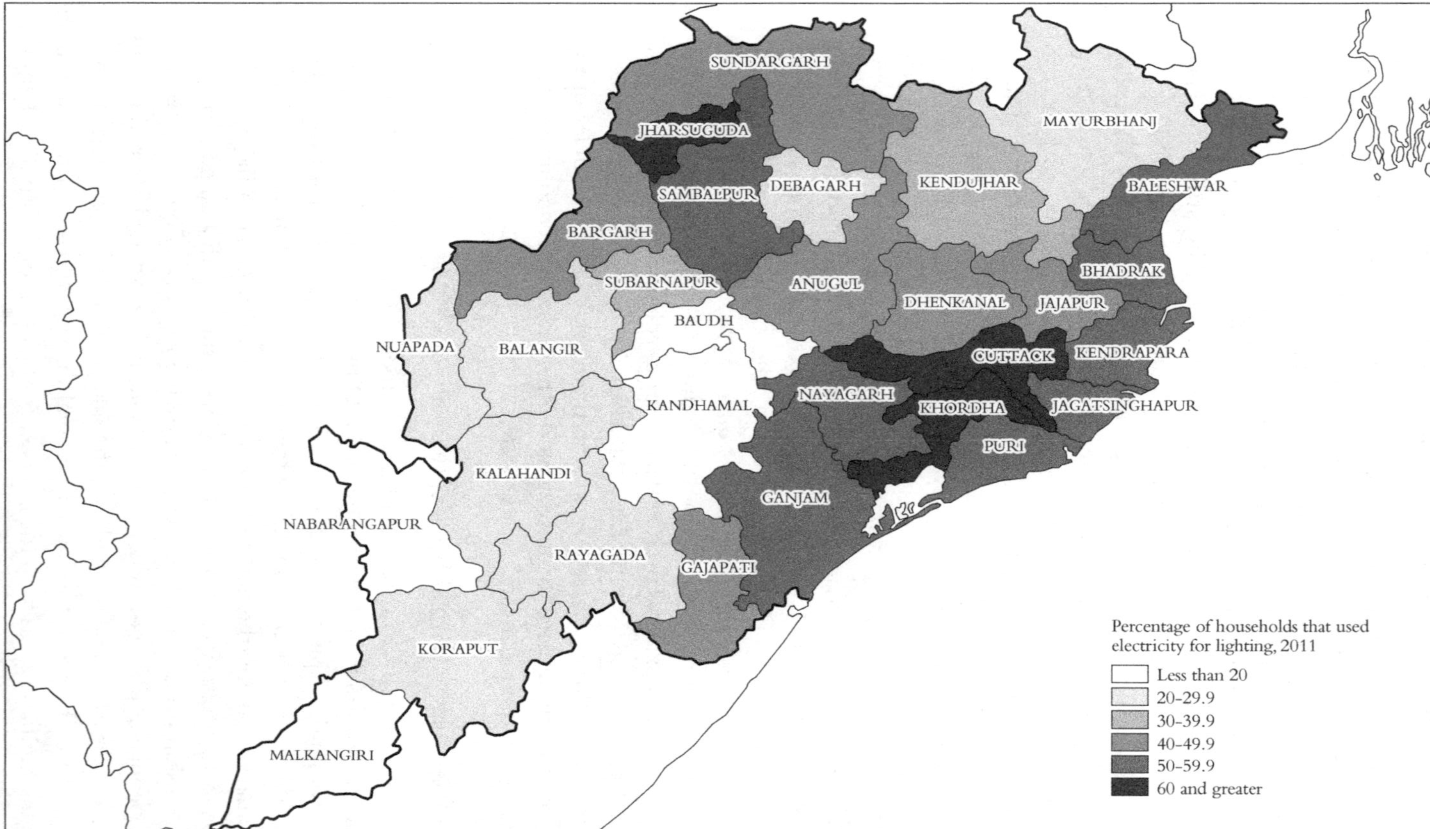

MAP 4.1 *District household electricity use for lighting in Odisha, 2011*
SOURCE: *Census of India, 2011.*

after independence, the two regions were markedly distinct in everything from patterns of land tenure, literacy rates, and social demography to the formation of political parties and histories of nationalist mobilization. And the consolidation of these territories within the single state of Odisha during the last sixty-odd years has not blunted these many differences (see Map 4.1).[5]

The language spoken by the majority of the population is Odia, although there are numerous regional variants, along with entirely distinct languages spoken among indigenous communities in the upland regions. Except for the states of the northeast, Odisha has the highest concentration of Scheduled Castes (SC) and Scheduled Tribes (ST) in India, who together make up nearly 40 percent of the state's population of close to 42 million.[6] The distribution of these historically marginalized communities is not uniform throughout Odisha however, with much higher proportions of SCs and STs in the former princely districts of inland Odisha, reaching 65 percent in Mayurbhanj District in the north, 45 percent in Sambalpur in the west, and 70 percent in Koraput in the south.[7]

In addition to the differences in social composition of the two regions, physical differences shape the economies of the two regions. Odisha has abundant mineral and mining wealth, the majority of which is concentrated in the inland districts.[8] The state's most industrially important natural resources are coal, bauxite, and iron ore. Bauxite, primarily used in the production of alumina, has been found in the inland regions of Bolangir, Sambalpur, and Kalahandi, all in the western hills.[9] Currently, the bulk of the bauxite is mined in Koraput, another inland district, by the National Aluminum Company Limited, which is owned by the central government and is India's largest exporter of bauxite.[10] Odisha and the state of Jharkhand in northern India have more than half of all India's coal reserves.[11] The possibilities for industrialization held out by these and other resources captured the imagination of Odisha's politicians, the most influential of whom hailed from the coastal plains.

Another factor that explains why successive governments privileged industrial development and neglected agriculture lies in the social basis of the state. Historically, Odisha's political elite have been largely disconnected from the countryside, with a social background among the upper castes and middle classes of urban, coastal Odisha. Unlike other parts of the country, in Odisha, the state and political elites throughout the postindependence period have had a much stronger relationship to urban and professional classes than to the state's many agricultural communities.

The sharp differences between the coastal districts and inland areas that spanned social composition, terrain, and economic base, were also reflected in electoral politics, particularly in the early decades after state formation. These political differences have their roots in the late colonial period, when the two regions had very different engagements with the independence movement and the subregional nationalist movement. In the 1910s and 1920s, student activists from the British regions were initially politicized through the Utkal Sammilani movement that demanded the unification of all Odia-speaking areas within the British territories, which at the time were distributed across the Madras, Bengal, Chhotanagpur, and central divisions.[12] Most of the movement's leaders came from the southern coastal areas ruled by the British and were of upper-caste backgrounds.[13] Rank-and-file support for the Utkal Sammilani came mainly from the middle classes in coastal Odisha, although the movement also drew some membership and support from the feudatory states. Typical of Anderson's "creole nationalists," the elites of British Odisha sought to parlay English-language educational opportunities into employment in the British administrative services, but found themselves disadvantaged by being linguistic minorities in their respective provinces.[14]

Although their cause was initially regional, many of these same leaders also became active in the national independence movement of the Congress Party. Along with movement activists from other regions of India, they pushed the national Congress leadership to accept linguistic identity as the organizing principle for the future of India.[15] When the Indian National Congress acceded to this demand in 1920, the fortunes and activities of Odia nationalism became more directly tied to those of the larger nationalist movement. Although the regional nationalist movement had drawn support from all Odia-speaking regions, including from the feudatory states, the new Congress nationalism that replaced it had a more antagonistic relationship with the latter.

The princely rulers in the upland regions served as intermediaries through whom the British collected taxes, which was in marked contrast to the practice of direct taxation in the coastal regions.[16] Relations between the royal families and the British crown were at times opposed, but during the nationalist period, the princes often cooperated with the British to suppress political dissent.[17] During the late colonial period, the Congress movement had helped to establish ancillary organizations, called Praja Mandals, in each of these feudatory states to serve as a counterpoint to the princes. The leaders of these political cells mostly hailed from middle-class families that had migrated inland from the coastal areas to seek expanded administrative opportunities.[18] In 1946, during the last set of elections before independence, the Congress bypassed the princes in favor of leaders from these new organizations, giving two ministerial posts to members of the Praja Mandals and thereby cementing the antagonism between the Congress and the princes.[19]

The lack of political cooperation between the Congress and the princely families continued after independence, when the coastal districts were merged with twenty-four of the twenty-six Odia-speaking princely states that covered the northern and northwestern parts of modern Odisha. In 1948, two ex-princes, R. N. Singh Deo and P. K. Deo, formed a new party to challenge Congress dominance.[20] The party, the Ganatantra Parishad (GP), was populated by royals and based almost exclusively in the former princely states. The GP's princely politicians took advantage of traditional social ties that continued to bind tribal and lower-caste populations of the former princely estates to their erstwhile rulers.[21] To further legitimate their rule, the princes also cultivated their *sub*regional identity, pointing to the historical, social, and economic differences between the coastal and inland areas to claim that the Congress Party, dominated as it was by coastal elites, was an ineffective representative organ for noncoastal citizens.[22]

In the first three elections to the legislative assembly of unified Odisha, held in 1951, 1957, and 1961, the electoral contest largely was a bipolar one between the Congress Party and the GP, with the former dominating in the coastal districts and the latter in the hills.[23] Both parties were populated by wealthier, more educated, and upper-caste politicians, but whereas ex-princes and large landholders populated the GP, it was professional politicians, administrators, businessmen, and teachers who predominated in the Congress.[24] Smaller socialist and Communist parties were active in the coastal areas of Odisha, but the Left parties never achieved in Odisha the kind of success they experienced in West Bengal or Kerala.[25]

In 1962, the GP merged with the Swatantra Party, then the second-largest party at the national level and India's main conservative party. The Swatantra Party represented a union of multiple political groups, including those coming from business, bureaucratic, and rural elite backgrounds. The incorporation of traditional rural elites was accomplished partly by way of co-opting regional associations of landed elites, of which the GP was a prime example.[26] In Odisha, the Swatantra and the Congress, whose leadership had by then changed hands to a younger generation, continued to be the two largest parties through the 1960s, each aligned at different moments with smaller parties formed by defectors from the Congress.

While waves of lower- and backward-caste political mobilization swept across southern, western, and northern India throughout the twentieth century, no comparable movements emerged in Odisha.[27] One explanation for this absence is located in the way that leaders of Odisha's scheduled castes and tribes have been politically incorporated. In the early elections, the GP—the party of the ex-princes—represented most constituencies. But the SC and ST politicians that emerged in those areas due to the system of reservations found their home in the Congress Party.[28] In the seven state elections from 1971 to 1995, 22 seats were reserved for SC candidates and between 33 and 34 for ST candidates in an assembly of between 140 and 147 members.[29] So although each community had a sizable presence in the state assembly, neither formed a cohesive political lobby. Instead, legislators from these communities tended to occupy subordinate positions within the mainstream political parties, initially the Congress and later dispersed across political parties.[30] In the past twenty years, the Bahujan Samaj Party (BSP) has consolidated its identity as the main vehicle for SC political aspirations in India. In its birthplace, Uttar Pradesh, it has had the most electoral success. But despite the large percentage of SCs in Odisha, there is only slight support for the BSP there.[31]

Just as there has been no successful autonomous political mobilization of SCs and STs, there likewise has been no political movement of middle-caste farmers in Odisha analogous to the backward-caste mobilization in the Hindi heartland through the 1980s and 1990s, or in earlier decades in southern and western India.[32] Partly, this was the product of how caste and class interact in Odisha. The Khandayats are largest peasant caste in Odisha.[33] The equivalents of the Khandayats in the states of Bihar and Maharashtra would be the Yadav and Maratha caste clusters, respectively, which spearheaded the non-Brahmin peasant-caste mobilizations in those two states. Given the political role of peasant castes in other states, we might have expected the Khandayats of Odisha to lead similar mobilizations. In Odisha, however, wealthier Khandayats tended to identify upward, with higher castes, while those less well-off invariably were grouped with lower-caste groups. By this reading, the Khandayats were thus divided along class lines, unable to serve as the basis of a broad-based, peasant-caste mobilization in Odisha like the ones that emerged elsewhere in India throughout the twentieth century.[34] Such class divisions could be overcome, as we saw in Maharashtra, but required the strategic cultivation of caste solidarity, a project that the Marathas began even before independence.

In explaining why peasant castes organized a vibrant non-Brahmin movement in Tamil Nadu but not in neighboring Kerala, Subramanian directs attention to variation in how caste groups related to each other.[35] In Tamil Nadu, Brahmins maintained a strict social and ritual distance from peasant castes, whereas in Kerala, Brahmins selectively intermarried with dominant peasant castes, which had the effect of collapsing the social distance between the groups and reducing the likelihood that the latter would array themselves against Brahmin dominance. A pattern similar to Kerala's may have obtained in Odisha as well. The 1921 census reports that intercaste marriage was "unheard of" in the province of Bihar and Orissa,[36] but ten years later, the 1931 census records an increasing tendency of Khandayats to intermarry with the upper-caste Karan community. Describing the growing laxity around endogamy, one census correspondent reported that although intercaste marriage continues to be socially condemned, the very power of that social condemnation has waned: "The man of power and pelf can shut the mouth of the caste people with gold and break the social rules with impunity."[37] The wealth of landowners was breaking down social barriers between wealthy peasant communities and upper castes. At the same time, Khandayats resisted the efforts of peasant and artisanal castes with less social and economic power, like the *chasas* of Puri district, to claim superior status.[38] So whereas the dominant landowning castes in today's Maharashtra and Tamil Nadu reached out and downward to increase their numbers, inspired by the ethical philosophies of Phule's *bahujan* (majority) in colonial Bombay, early non-Brahmin activism in colonial Madras, and a tactical understanding of the power of numbers in a democracy, in Odisha, as in Kerala, the dominant peasant castes reached upward through intermarriage.

Large landowners in inland Odisha—the former feudal families—were similarly unable to seize and hold on to state power. In 1971, the national government passed a constitutional amendment abolishing the hereditary, financial privileges that the Indian government had granted to the royal families as part of the accession agreements of the princely states to the Indian union. With this, the Congress Party at the center ensured the eventual dissolution of separate princely political formations, and what had been a fairly sharp political split between coastal and inland Odisha began to fade.[39] The elections held in 1974 were the last in which the Swatantra Party made a strong showing, after which Odisha had no political party with an exclusive base among the feudal families. Instead, the two political coalitions that have dominated Odisha's politics from the late 1970s onward—the Congress and its various offshoots on the one hand, and the various incarnations of the Janata Dal on the other—have had both similar social bases and strikingly similar stances toward economic development, favoring heavy industry and the development of the mining sector.

Odisha's Development Goal of Extractive Industrialization

There is a tendency to date the exuberant support for extractive industry to the period of neoliberal economic policies. This is particularly true in the resource-rich areas of eastern and central India that are home to the majority of India's indigenous populations. As one author puts it with reference to these mineral rich districts: "In recent years the alliances of neoliberal state policies and multinational corporate interests have sought to harvest these riches to fulfill the dream of 'India Shining.'"[40] Often described as a tidal wave of accumulation by dispossession, these patterns of land acquisition for industrial, mining, and large-scale development processes are considered emblematic of India's liberalized and globalized economy.[41] While land dispossession is undoubtedly a feature of the contemporary political landscape, it is not unique to the current period. The histories of antidevelopment struggles that began in Odisha in opposition to the Hirakud Dam in the 1940s are one important testament to this.[42] The idea of natural resource extraction as the basis of industrialization is likewise not new. Indeed, Odisha's particular development regime has been configured around an aspiration to maximize its resource wealth since before independence. An emphasis on industrialization as the surest route to development, with little attention for agriculture, has remained a constant feature of the development regimes of Odisha's various parties, factions, and coalitions from independence to the present.[43]

As one activist and writer in the state described the political milieu in the past twenty years, "politics in Odisha is like a *jatra* performance."[44] He was referencing the popular folk-theatrical tradition found in eastern India and Bangladesh. Although the actors playing one role or the other periodically change, at the end of the performance, it is clear they are all from the same tight-knit theater troupe. The political spectrum, to him, is similar; it is full of political parties that outdo themselves during election campaigns to differentiate their ideology and mission from one and other, but once in office pursue policies that are indistinguishable one from the next—equally apathetic to agricultural interests and eager to industrialize. This remained true even during the 1990s and afterward, when, as one newspaper editorial put it, an emphasis on heavy manufacturing and industry as the instruments of development already had become "anachronistic."[45] By then, Odisha's neighbors had already launched ambitious efforts to cultivate the technology-based service industries that would emerge as the engines of economic growth in the 2000s.

As I argue in previous chapters, despite the centralizing tendencies of the industrial planning regime, India's subnational units were given a unique set of tools to shape the course of economic development in their territories. Infrastructural development—transport, communications, and electricity—has been among the provincial state's most potent tools. We can see the material consequences of Odisha's emphasis on industrialization in a spatial pattern of infrastructural development that largely bypassed rural areas and was concentrated in the coastal districts of the state, and in the allocation of scarce developmental resources to industrial rather than agricultural uses. The main beneficiaries were central government–owned enterprises, state government–owned units, and regional capitalist firms with close ties to the state.

Much scholarship on the political economy of Indian industrialization examines the relationship of the national business class to the central state's efforts at industrial transformation. We have comparatively few systematic and empirically grounded analyses of the relationship of provincial states to regional capitalist classes. But what little work has been done shows the extent to which local circuits of capital have informed regional political economies.[46] In Odisha, the close working relationship of an industrial class and a provincial state was not sufficient to ensure the kind of industrialization that would either structurally transform the domestic economy or raise incomes and living conditions for the state's citizens. So while there is sufficient evidence of the state's attempt to foment industrial transformation, these efforts were never enough to transform the state into a dynamic industrial center (see Table 4.1). Odisha remains a poor state in terms of per capita income, and one in which the vast majority of the population continues to rely on agriculture for its livelihood.

TABLE 4.1

Average annual growth rates in Odisha's industrial sector

Period	*Number of factories*	*Number of employees*	*Value added*
1960s	3.61	5.98	23.45
1970s	4.25	2.05	21.05
1980s	2.60	3.22	16.38

Source: Vinod Vyasulu and A. V. Arun Kumar, "Industrialisation in Orissa: Trends and Structure," *Economic and Political Weekly* 32, no. 22 (May 31, 1997): M46–53.

The dilatory growth of Odisha's industrial sector, as evidenced through the quantum and value of manufacturing units, is also reflected in the sectoral composition of the state's economy. Why, despite the efforts of the state government, did Odisha fail to industrialize? Although there remains disagreement about precisely why some parts of India industrialized relatively more successfully than others, one strain of the literature argues that the postindependence regions with the most dynamic economies were those in which prior investments had raised agricultural productivity.[47] The resulting agricultural surplus served as a basis from which emerged diversified agro-industrial activity, often with the aid of the local and provincial state. Examples of such a model are Punjab and Maharashtra, where postindependence generations of regional industrialists and agro-industrialists expanded from the countryside with close ties to the state. In Odisha, by contrast, the industrial class emerged primarily from professional and urban contexts and relied on education and political connections to facilitate their entrepreneurial activities. Their abilities to expand, though, were limited by insufficient capital sources. Sinha has convincingly shown how, even during the height of central planning, state governments were nevertheless able to intervene in industrialization to great effect. In that study, although the states that successfully industrialized—Gujarat and, to a lesser extent, Tamil Nadu—had robust agricultural sectors, this agrarian foundation per se is not a part of Sinha's explanation for successful industrialization.[48]

Another possible explanation for the failure of industrial transformation in Odisha, despite tight state-business relations, might point in another direction. One cross-national comparative study of developmental success in Korea and Taiwan looks to their historical antecedents. Through a comparative analysis of those two successful cases and two relative developmental failures in Latin America, Davis proposes that a foundation of small to medium-sized farmers was a necessary political base out of which the successful developmental regimes of East Asia emerged in the 1970s and 1980s.[49] Rural middle-class incorporation gave those two states both the necessary leverage to maintain autonomy from industrial capital, even while engaged in supporting it actively, and also imbued development with an ethos of discipline. In Odisha, we might speculate that the absence of a rural middle-class base helps to explain why the state was supportive of industry but could not play a disciplinary role. That is, the state often provided support in the form of subsidized inputs, like electricity and finance, but was unable to enforce the efficient use of those resources. The consequence in many cases was that the state later stepped in as owner of firms that would otherwise fail entirely, a task that was common to the public sector in other parts of India as well.

A third explanation for uneven industrialization points to the nature of fiscal federalism in India, particularly during the first forty years of planned development. Under the centrally planned regime, investments were meant to be spread across regions to secure balanced growth. This was, in fact, one of the main justifications for the labyrinthine licensing regime that was a hallmark of India as a centrally planned economy. Despite the intent of spreading industrial growth, a plethora of analyses from both scholars as well as the government's own inquiry committees document the lopsidedness of industrial investment during India's first four decades.[50] Many states in India complained of "step-motherly" treatment from the central government; Odisha was no exception.[51] From the late 1960s onward, a political bargaining model seemed to obtain in the allocation of central investments and industrial licenses. Within this framework of a dominant center and centralized planning, states with greater political leverage (which includes those with larger populations, those whose state governments had partisan affiliations to the government in New Delhi, and those well represented in the cabinet) exercised more influence over investment decisions.[52] Odisha's relatively small population meant that the state accounted for a small parliamentary contingent and could not wield the political clout of more-populous states like Uttar Pradesh, Maharashtra, and Tamil Nadu. But with its small size and relatively less influence, in some instances, politicians from Odisha nevertheless succeeded in using social and political connections to channel resources from the central government to the state, as with Biju Patnaik.

Electricity Infrastructure and Development in the 1960s

Although the state in Odisha failed to achieve a dramatic industrial transformation, there is nevertheless a consistent pattern of state infrastructure support to industry. The imprimatur of this policy was stamped in the 1960s, and its legacies persisted in subsequent decades. As discussed previously, during these early years of state formation, the Congress Party drew electoral support primarily from the coastal districts, with an organized political opposition that gathered its membership from royal families and its electoral support from the erstwhile feudatory estates. Throughout this decade, various state agencies were formed to facilitate industrialization, with little corollary effort expended on agricultural development. Among other strategies, chief ministers lobbied the central government on behalf of local industry, provided financing to facilitate industrial expansion, and, where necessary, nationalized failing industrial enterprises through the state's industrial corporation.

The state's planning agencies and government ministries also privileged industry in allocating scarce infrastructural inputs. Indeed, the lion's share of electricity resources was put toward industrial expansion. The extent to which the state's infrastructural development bypassed agricultural in favor of industry is unique among the cases considered in this book, and in India as a whole. In Odisha, rich in coal deposits and in potential for hydroelectric development, the projected gains from new endeavors that exploited these resources were almost entirely set aside in advance for industrial use.

In many ways, newly formed Odisha was merely following patterns that had been established in the previous decade, when Odisha's vast Mahanadi River valley became the site of one of India's first three multiuse river projects. Like the Damodar Valley Authority (DVA), the Mahanadi River project was intended to provide flood control in the river's delta regions, divert water for upgraded canal systems on either side of the river's lower reaches, and serve as one of the state's first substantial sources of electricity through hydroelectric dams built at intervals along the river. As one editorial extolled, "instead of floods, death and devastation there will be perennial irrigation, bigger food crops and cheap electricity to raise the standard of life in the villages."[53] The editorial echoes the discussion from Chapter 2 of the imagined unifying role that electrification would play in modern India, bringing a resource that had previously been confined to urban spaces also to rural India and, in the process, fulfilling another of modernity's technological promises—to master and harness the destructive power of nature. But the outcomes of the Mahanadi project in Odisha belied this expectation.

Prime Minister Nehru traveled to Odisha in 1948, when the Hirakud construction began, and returned there just over ten years later to flip the switch on the new hydroelectric supply that would augment Odisha's meager capacity of 4,845 kW by 120,000 kW. Several years before the electricity started flowing, though, the state had appointed a committee headed by the finance minister to determine the rates at which electricity from Hirakud would be sold to various classes of consumers and to reserve capacity for consumers with whom it negotiated contracts. The first major consumer of Hirakud's electricity was the central government's own Rourkela steel plant. Other large industrial consumers and the amount of energy they would require from the plant include an aluminum production plant (25,000 kW), a ferro-manganese plant (50,000 kW), and the Jharsuguda Defense Chemical Factory (15,000 kW).[54] Given the vast quantities that were set aside for these large industries, very little remained for rural consumers, either for smaller-scale industrial units in the textile and agro-processing sectors or for farmers. The same was true of the 30,000 kW that was Odisha's share from its other large-scale hydroelectric project on the Machkund River, along the state's southern border with Andhra Pradesh. Unlike Odisha, Andhra Pradesh used the resultant electricity for multiple purposes, including increasing agriculture productivity through electrification of irrigation pumps.

Though the supply of electricity grew from these two hydroelectric and later thermal projects, the amount of electricity consumed by the agricultural sector remained negligible. Rural electrification was a slow process elsewhere, made especially difficult in the hilly terrain of western Odisha, but the pace in Odisha was slower than in any other part of India outside of the northeast. In 1960, Odisha's industrial connected load made up 89.3 percent of the state's total connected load, whereas the connected load for irrigation use was 0.14 percent. The only states that had a more skewed ratio were West Bengal and Assam. The sales of electricity from this period reflect the same bias toward industrial consumption.[55]

In the third plan period (1961–65), which was the first period after the OSEB's formation in 1961, Odisha set lower targets and committed fewer resources to its rural electrification program than most other states.[56] Odisha planned to electrify 165 villages (0.34 percent of the total villages in Odisha, according to the 1951 census) and allocated Rs. 14 million, whereas Andhra Pradesh set its target at 925 villages (3.5 percent of all villages) and allocated Rs. 90 million. Odisha's restrained approach to rural electrification continued in subsequent plan periods. Even during the period of the Green Revolution—from roughly 1965 to 1975, when New Delhi renewed its emphasis on rural electrification—the gains in Odisha were nominal.

The language of the plan documents from the early 1960s also illustrates the state's heavy emphasis on industrialization and the relative absence of the rural from the imagined future of modern Odisha. As discussed in Chapter 3, Maharashtra's plan documents from the same period repeatedly stress the goal of a "modern agro-industrial society." Odisha's planners took a different view, emphasizing instead that the state must use its mineral wealth to support industrialization. The preface to one plan document records that "paradoxical as it is—Orissa stands industrially backwardmost [*sic*] amidst her bountiful potential mineral, land, and water resources. With these resources systematically exploited she is assured of industrial and economic advancement within a decade of planned development."[57] A. N. Khosla, governor of the state in 1962, authored an ambitious master plan for the state that takes a similar view. When Nehru appointed Khosla as governor in 1962, the latter was already famous as an irrigation engineer, having worked in both colonial and postindependence central government bureaucracies for several decades. In a master plan for the integrated development of the state's main river basin, originally given as the Visvesvaraya lecture for India's Institution of Engineers, Khosla prophesied that "Orissa is destined to be the most industrially advanced State in India."[58] The main obstacles standing in the way of this were the lack of infrastructure, chief among which was electricity. The master plan to develop a series of hydroelectric dams on the Mahanadi River was meant to resolve this deficiency.

Plan expenditures during the 1960s reflect the state's commitment toward rectifying the deficiencies that Khosla had identified. During the first four years of the *Third Five-Year Plan* (from 1960–61 to 1964–65), total plan expenditures came to 166 crores, out of which the largest share, 36 crores or 22 percent, was spent on power. In privileging the development of power, Odisha was similar to other states. Within the allocation of funding for power, however, Odisha set aside a much smaller percentage to rural electrification than other states.

During the 1960s, while the planning bodies were thus allocating financial resources to rectify the deficiencies that the planners had pointed out, the government also expended considerable resources to build a set of institutions that were expected to serve as the foundation of industrial development. The state adopted a new, aggressive stance toward industrialization during the early years of the *Third Five-Year Plan*. As the state's Planning Commission puts it on the very first page of one of its plan documents, "it was decided that the State should play a more direct and active role than hitherto in regard to establishment of industries."[59] In addition to the OSEB, which was established in 1960, the state set up an Industrial Finance Corporation, the Industrial Development Corporation, the Orissa Mining Corporation, and the Orissa State Commercial Transport Corporation. Alongside these, the government also laid plans to build a new port at Paradeep, the Balimela Dam, and the Express Highway in the first half of the 1960s.

The Congress-ruled state during the 1960s also established a close working relationship to the nascent regional capitalist class, one that was sometimes muddied by the crossing back and forth of individuals from private industry to the government. Throughout this decade, this closeness provoked opposition politicians in the GP to accuse the government of allowing individuals to benefit privately from public resources, an accusation that translated into demands that the central government intervene. After resisting these requests for several years, the central government finally complied in 1967 by appointing a judicial inquiry, the Khanna Commission, which submitted an exhaustive report of more than 800 pages in early 1969.[60] The report excavates seventy allegations of impropriety against fifteen politicians who variously held the posts of chief minister, cabinet minister, and deputy minister in the Congress-led government from 1961 to 1967. As a historical document, the commission's report suggests two things: first, the state government actively intervened to support industrialization through the vehicle of a regional capitalist class; and, second, local industrialists had significant influence within the state.

At the heart of the Khanna Commission's investigation was Biju Patnaik, one of Odisha's most formidable politicians. After an early career as a pilot during World War II and heroic exploits during the tail end of the nationalist movement, Patnaik's first forays in public life were not as a politician but as a diversified entrepreneur, with interests spanning airlines, textiles, refrigerators, mining, iron ore production, and industrial manufacturing. After he moved into politics, Patnaik served as a member of the state legislature, twice as chief minister (1961–63 and 1991–95), and as a cabinet minister at the national level. During his first stretch as chief minister, it was the close relationships of his private firms with state-owned financial and industrial development initiatives that drew opposition protest. Firms owned by Patnaik and his family, as well as the families of other ministers in his government were implicated in the investigation. Patnaik himself was accused in thirty-seven charges; his successor to the chief minister's office, Biren Mitra, who was deputy chief minister initially, was accused of forty-two. The allegations range from the channeling of public funds through private companies to benefit Congress campaigns to favoritism in awarding contracts for state projects like the port at Paradeep. Patnaik's stature in the state was such that even after resigning from his first stretch as chief minister in 1963, he continued to serve as the chairman of the State Planning Board, and worked closely with his successor to continue the program of industrialization.

The complex connections between the state and business interests were not exclusive to the Patnaik family, as the full text of the Khanna Commission report makes plain. The deputy chief minister, Biren Mitra, was also engulfed in controversy for his role in Patnaik's Kalinga Industries, as well as the success of his wife's commercial ventures in securing government contracts, particularly during his tenure.[61] In the end, the commission declared that Patnaik, Mitra, and the others were guilty of "various acts of impropriety," but since there was little evidence of direct financial benefits to any of them, there were no further criminal proceedings. Although the evidence of direct financial gain might have been missing, the investigation firmly established the extensive ties linking Odisha's politicians through their families and holding companies to the state apparatus through industrial holding companies, state finance corporations, and government contracts and procurement.

The political fallout of the corruption scandals that tarred the Congress Party was an old-fashioned thumping during the 1967 assembly elections. The Congress Party lost decisively, and a new coalition formed between the Swatantra—which in Odisha was almost exclusively the party of the princes—and a new faction that broke away from the Congress, called the Jana Congress. This Jana Congress-Swatantra government, led by R. N. Singh Deo as chief minister, represents one of the few times that the state's development agenda seemed to shift in favor of large landowners and agrarian interests. The government was short-lived, however, and beginning in 1971 and extending for the next decade, politics in Odisha was characterized by unstable coalitions of disparate groups and breakaway factions of the Congress Party. One manifestation of the remarkable instability of politics during the 1970s was New Delhi's imposition of "president's rule" in the state in 1971, 1973, 1976, and 1977.[62] The apparent instability, however, contrasts with enduring features of the state's development agenda throughout this period, which coupled a prioritizing of industrialization with a neglect of agriculture.

1970s–1980s: Odisha During the Green Revolution

In the face of persistent food shortages during the early and mid-1960s, the national government of India introduced a set of policies to spur agricultural development through more-intensive cultivation, dubbed India's Green Revolution. Although the Green Revolution is often generalized as a national phenomenon, only certain pockets of agriculture underwent anything like the dramatic changes usually indexed by the word *revolution*. There were very few parts of Odisha that experienced yield increases like those obtained in wheat cultivation in Punjab or rice farming in Andhra Pradesh. Those success stories of the Green Revolution took place in states that had a clear public commitment to increase agricultural productivity.

In the effort to extend the gains of the Green Revolution outside of its early core areas, New Delhi created incentives to spread the necessary technologies more widely across rural areas, but these too had uneven effects. Among these was the creation of a new Rural Electrification Corporation (REC) to help subsidize the expansion of what had already emerged as a vital input for farmers, allowing them to use electric tube wells to pump groundwater resources for irrigation. But these new pockets of public finance were also unevenly spread. As a 1975 report by Odisha's Planning and Co-ordination Department shows, the central government's push to support electrification had only limited results in Odisha, which continued to exhibit patterns of electricity consumption that favored industry and urban coastal districts.

According to the 1971 census, 92 percent of Odisha's population lived in approximately 47,000 villages.[63] As of 1969, only 967 villages were electrified in the state. Two years later, this number had doubled, but still the percentage of Odisha's villages with electricity was only 4 percent of the total. The national average for rural electrification in 1971 was 19 percent, so the base from which the renewed electrification program began was already substantially lower in Odisha than elsewhere. Over the next five years, with support for forty-four separate rural electrification projects from the REC, the percentage of villages electrified in Odisha increased to nearly 25 percent, still low compared to the national average.

Electrification was also regionally uneven. By the end of 1974, more than one-third of the villages in the coastal districts of Cuttack, Puri, and Ganjam were electrified, but inland districts like Phulbani, Kalahandi, and Koraput had village electrification rates closer to 5 percent (see Table 4.2).

The figures on village electrification also did not tell the full story. A sample survey of villages across districts found that out of seventy-seven electrified villages, thirteen did not have a single consumer of electricity. In those cases, the utility brought the transmission lines to the village boundary, but not into the village itself.[64] Given the way electrification was enumerated, the utility could still count such villages as electrified, even though neither homes nor fields used electricity.

Rural electricity is used for multiple purposes including street lighting, household consumption, small and medium agro-industry, and irrigation. Relatively few villages had streetlights at this time, and the rural industries that made use of electricity in Odisha were mostly rice hullers, whose need was confined to certain times of the year. The largest use of electricity was for irrigation purposes. To get a sense of the productive ways that electricity

TABLE 4.2

District rural electrification in Odisha as of 1974

District name	*Total rural population*	*Electrified villages by end of 1974–1975 (%)*	*Rural population residing in electrified villages (%)*	*Number of electrified irrigation pumps*
Balangir	1,176,994	19.38	26.81	230
Balasore	1,730,350	28.07	33.63	288
Cuttack	3,522,055	44.45	50.68	993
Dhenkanal	1,242,102	34.46	55.50	214
Ganjam	2,033,952	34.78	62.66	552
Kalahandi	1,107,316	6.33	12.99	63
Keonjhar	888,167	25.68	39.08	59
Koraput	1,876,022	8.39	15.22	253
Mayurbhanj	1,394,249	18.16	26.23	73
Phulbani	602,107	1.59	7.03	13
Puri	2,111,712	37.06	42.68	85
Sambalpur	1,623,121	22.33	37.11	60
Sundargarh	791,073	40.16	55.72	82
Odisha Total	20,099,220	24.52	38.69	2,965

Source: State Evaluation Organisation, Planning and Coordination Department, *Report on the Evaluation of Rural Electrification Programme in Orissa* (Bhubaneswar: Government of Orissa, 1982). Data assembled from pages 3–7, tables 1–4.

was put to use, Table 4.2 gives information on the number of electrified lift irrigation points in each district. For both rural households with private pump sets and those with access to publicly owned lift irrigation points, the gains to agricultural productivity were enormous. The net return per acre of farmland using electrified pump sets was three times that of farms without irrigation.[65] There were similarly large gains for rural industries that used electric as opposed to diesel motors.[66] In addition to the economic gains from electricity, the sample survey asked villagers whether electricity had transformed their lives. The results were overwhelmingly positive: more than three-quarters reported that electricity lowered the costs of lighting, reduced household fires (many of which were started because of oil lamps), improved their children's study habits, and allowed household chores to be done after nightfall.[67]

Despite the evident benefits from increasing electricity access, during the 1970s, while states like Maharashtra were pouring resources into rural electrification, Odisha instead continued and even expanded its subsidies for industrial consumers as a way to attract industries to the state. The OSEB's annual reports were explicit in so justifying the state's extremely low tariffs for industrial consumers: "In its attempt to encourage such [power intensive] industries, the Board, in the recent revised tariff enforced from April '75, has prescribed a fairly low tariff in comparison with other States in India . . . It is hoped that prospective entrepreneurs will take advantage of these facilities and come forward in large numbers so nature's gift can be best utilized for the benefit of the country as a whole."[68] The conscious strategy was to use low electricity tariffs to compete with other states for industrial investors. As we see in Chapter 2, during the 1960s, central government agencies, power sector analysts, and the World Bank had blamed this very tendency for persistent problems in India's electricity sector. The subsidized rates for industrial consumers consistently and negatively impacted the OSEB's financial position during the 1970s, which further eroded the board's ability to finance rural electrification projects. The utility's financial losses for the years 1971–72, 1972–73, 1973–74, and 1974–75, were respectively Rs. 90, 190, 320, and 410 lakhs due to rural electrification, and Rs. 200, 260, 387, and 770 because of industrial subsidies. In each of these four years, the losses from industrial subsidies were higher than those for rural electrification, and in two years, industrial subsidies cost twice as much as rural electrification.[69]

The government's neglect of rural development continued into the next two decades. By 1983, only 43 percent of villages in Odisha were electrified.[70] The only states with a worse record were Bihar (with 34 percent) and Assam (with 31 percent). By 1991, Odisha recorded 67 percent village electrification.[71] In contrast, Maharashtra, Andhra Pradesh, Gujarat, Himachal Pradesh, Punjab, and Haryana all claimed 100 percent village electrification.[72] Whereas subsidies to agriculture were increasingly blamed for the poor fiscal health of finances in these states, Odisha was largely free of this problem. Although farmers in Odisha were charged low tariffs, as they were in most other states, the meager rural electrification program of earlier decades ensured that the amount of electricity consumed by farmers was negligible. In Odisha, farmers used only 3.8 percent of all electricity in the state in 1990–91, when the national average was between 30 and 40 percent.[73] By 1994–95, agricultural consumption had grown to only 5.1 percent of total electricity consumed in Odisha.

Instead of a robust program to enhance agricultural productivity, which was part of the Green Revolution policies in many other parts of the country, and faced with growing capital scarcity at the local and provincial levels, in Odisha the state's attention during these decades seemed to be ever more directed toward lobbying the center for public investments in the power, iron, and steel sectors. During the 1980s, the Congress returned to power in

Odisha, this time under the leadership of another Patnaik, J. B. Patnaik.[74] During the 1980 election, when candidates in neighboring states, like Andhra Pradesh, were promising subsidized electricity for farmers and subsidized grain for consumers, J. B. Patnaik's campaign promised "1000 industries with 1000 crore in 1000 days."

Many analyses of why the promised industrial growth did not materialize emphasized a lack of electrical power.[75] There was indeed a severe electricity shortage in Odisha in 1982, when low reservoir levels restricted the output of the state's hydroelectric stations. Odisha's local business groups complained that after having enticed entrepreneurs to the state when he first came to office, J. B. Patnaik had betrayed the interests of industry. They complained of delays in project financing and raw materials, and primarily of the recurring power reductions that cut deeply into their profitability.[76] Although the number of industrial firms did increase during the 1980s, industrial growth in terms of the value added to the state's gross domestic product actually decreased in the 1980s relative to earlier decades. The electricity shortages of the 1980s, and the impediment they posed for industrialization, form an important context in which to understand the events that followed in the 1990s, when Odisha became the first state in India to follow the globally dominant model of utility privatization.

Odisha in the Era of Market Reforms

In elections that were held in 1990, Biju Patnaik, who was by then arguably the most influential leader in Odisha's postcolonial history, returned with an overwhelming majority to the chief ministerial office. In his second stint as chief minister, Patnaik served a full five-year term. Not only was this a complete term, but it can be considered a more successful one in that his efforts to promote industrialization were more effective. From his early promotion of private entrepreneurship allied to the provincial state, to a commitment to state-led development during the 1970s and 1980s, by 1990, Patnaik had returned full-circle—frustrated that forty years of central planning had not benefited Odisha and newly energized by the possibilities of private capital-led industrialization. During the late 1970s, Biju Patnaik had moved from state politics to the central government, successfully channeling central public investments to his home state. He served as cabinet minister in the two Janata national governments that followed Indira Gandhi's "Emergency," first in the role of minister for steel and mines from 1977 to 1979, and then minister for steel, mines, and coal from July 1979 to January

1980.[77] As a union minister, Biju Patnaik had brokered a deal for investment in an aluminum plant, construction for which began in 1981.[78] His return to state politics in the 1990s coincided with India's general turn toward the market, when states now entered the new context of competitive federalism.

Biju Patnaik's election campaign in 1989 consisted of three major planks: a new steel plant in the state, reduction in farmers' loan burdens, and an anticorruption sweep in the government and bureaucracy.[79] Politically, Biju Patnaik was initially in a strong position to carry out his election promises. His party had won the elections with an enormous majority, a victory that the media attributed to Patnaik's extraordinary charisma.[80] Out of the 147 legislative seats, politicians from Patnaik's party, Janata Dal (JD), contested 139 and won 123, easily displacing the Congress government that had held the executive branch previously. The JD's victory was not so overwhelming, however, when judged by actual votes; 54 percent of the votes cast went to the JD, whereas its main opposition party, Congress, won roughly 30 percent of the votes but only ten seats.[81]

This election outcome strengthened Patnaik's hand, shoring up his political capital at a critical moment. One of Patnaik's fellow party members stated: "Biju-baba is the heart and soul of this government. The remaining could at best be termed ringside spectators."[82] That Patnaik held a similar view of his importance in the party and government became clear by the way he formed his government. In addition to occupying the chief minister's post, Patnaik also reserved for himself several other cabinet ministries, including home, general administration, industries, mining and geology, and planning and coordination positions. The list itself is suggestive of Patnaik's priorities, which included anything related to the industrial development of the state. To give institutional fillip to his plans for industrialization, Patnaik created a new department of steel and mines, the directorship of which he also held. By creating the first such state-level ministry in India, Patnaik was attempting to devolve the authority of the central government to the state level.[83]

During a speech given early in his term, Patnaik expressed dissatisfaction that during the previous forty years, Odisha had been able to neither "improve the relative position vis-à-vis other states in the matrix of economic development" nor change the "pattern of intersectoral contribution to the net domestic product over the years."[84] That is, Odisha remained an overwhelmingly agrarian state. The intentions of neither Odisha's earlier development plans nor central planning for regionally balanced development had been able to alter this fact. The share of the primary sector in the net state domestic product (NSDP) fell from 55.09 percent in the early 1980s, to 47.04 percent one decade later, and to 39.23 percent at the end of the 1990s.[85] The population of the state, however, continued to be overwhelmingly rural and tied to the land.[86] Relative to the rest of India, Odisha had extremely low levels of industrialization. In a ranking of the fifteen largest states in the country in 1986–87, Odisha occupied the fifteenth position in terms of the percentage of value added by its factory sector.[87] This was roughly the same position Odisha occupied in the preceding three decades.

Foreshadowing the sharp turn that Odisha would make toward the private sector, in the same speech, Patnaik also complained of "dilatory bureaucratic fetters which stifle initiative and growth."[88] And in a later interview, he condemned the inefficiencies of the public sector, which "employ five men where one is required."[89] He also betrayed a ruthless streak, once asserting that "no development is possible without human sacrifice. Be it Ashoka or Peter the Great, human lives perished but that is how their great empires flourished . . . So why do we shy away from paying a price for development?"[90]

Patnaik and other state leaders diagnosed capital scarcity to be Odisha's main obstacle to rapid industrialization. When the central government opened up industrial sectors to both indigenous and multinational capital, state politicians saw a new opportunity. Privatizing electricity—one of the most significant economic inputs for industrial activity—became useful both as a way to remove an industrial bottleneck and also as a way to signal the state's commitment to economic liberalization by vacating industrial space to the private sector. S. K. Mohapatra, who had risen through the ranks of the OSEB to finally serve as chairman of the utility in the early 1990s, when the government first initiated privatization, suggested that Biju Patnaik was convinced of the greater efficacy of a privately owned distribution system. Patnaik was also aware, according to Mohapatra, that giving Odisha the status of the first privatizer would dovetail nicely with the emerging pro-business stance that Patnaik wanted the state government to adopt.[91]

Foreign Direct Investment in Odisha

The government of Biju Patnaik that came to power in 1991 in Odisha inherited a state with severely restricted public investment capacity. But the combination of abundant natural resources and central government policies that were meant to encourage private investment ensured that Odisha emerged as a prime target of private interest, particularly for large-scale projects. While the total number of proposed investments was much larger in other states, particularly in those more industrially advanced, the average size of proposed investments was largest in Odisha.[92] Judged both on a per capita basis and as a percentage of the state's gross domestic product, these investment proposals were also far more significant in the poor state of Odisha than in other larger and more developed states. Odisha's share in total investments in India also increased substantially from the pre-reform period to the post-reform period. The increase was even sharper for heavy industries, which includes manufacture of nonmetallic minerals, metals, metal products, machinery, and transport equipment.[93]

In 1990 and 1991, Odisha's mineral and fossil fuel resources in particular were garnering international attention. Both Indian and foreign firms proposed a variety of projects in the state's mineral sector. One such investment possibility was a joint-venture project between the state government of Odisha and the Jindal group of industries, a prominent Indian industrial house. The joint-venture project was intended to fulfill an election promise.[94] After that plant failed to materialize, Odisha formed another joint venture, this time with UK-based Caparo Industries to build a steel plant at Duburi, in the Jaipur District.[95] The Indian public-sector company, National Aluminum Company, intended to build an export-oriented alumina refinery in tandem with Hydro Aluminum of Norway, to be located in the southern district of Koraput.[96] In the same district (where the bulk of the state's bauxite is found), two of India's leading private industrial houses, Indal (part of the Aditya Birla Group) and Tata Industries, planned to build an alumina refinery.[97] Similarly, Odisha's chrome ore deposits,[98] discoveries of platinum,[99] and large reserves of granite[100] also generated a slew of proposed investments.

The sector that received the most investor attention was the electricity generation sector. Odisha's massive coal reserves were emerging in this period as the cornerstone of the Indian government's plans to augment generation capacity. By the 1990s, the integration of India's transmission system was well under way. Once complete, it would be possible for electricity to be generated close to coal mine pitheads and other fuel sources—and then efficiently carried to the industrial and urban load centers via high-tension wires that created a unified national grid.[101]

The Central Electricity Authority (CEA) planned to develop fourteen coastal thermal power plants to be built and operated by the National Thermal

Power Corporation (NTPC). These were to be considerably larger than existing plants, to benefit from economies of scale.[102] The first five of these large-scale power projects were all to be sited in South India, and the bulk of the coal to fuel these plants was to be taken from the coal fields in Talcher in southeastern Odisha and from Ib Valley in northern Odisha.[103] A joint venture tentatively proposed to export coal to France in exchange for joint-venture support from Charbonnage de France to Mahanadi Coalfields Ltd., the Odisha-based subsidiary of Coal India, the public-sector company in charge of all coal mining in the country.[104]

Several other Indian and foreign firms, many of them American, initiated plans to build coal-fired power plants in the state. Southern Electric International (SEI), a subsidiary of the Atlanta-based Southern Company, started negotiations with the Odisha government in 1991, kick-started by a visit from Chief Minister Biju Patnaik to the company's headquarters.[105] SEI planned to build a Rs. 60 billion ($2.3 billion) power plant of 2,020 MW in the Ib Valley, one of Odisha's two major coal sites, located in the far northwestern districts of Jharsugada and Sundargarh. The size and cost of the project, which the US company was to build, own, and operate, were unprecedented in Odisha. A preliminary Memorandum of Understanding (MOU) was signed between the vice-president of SEI and Prime Minister P.V. Narasimha Rao in December 1991.[106] A second US firm, Northeast Energy, wanted to build a 500 MW plant located in the eastern Cuttack District, at a cost of Rs. 15 billion ($576.9 million).[107] In 1992, an American firm, International Contracting and Marketing Corporation, based in New York, was putting together a consortium of US firms to build several power plants in India, among them a coal-based plant in Odisha. This was partly a product of the efforts of both the US and Indian governments to encourage American investments in the Indian energy sector. In June 1992, an Indian delegation traveled to the United States to encourage potential investors; this visit was reciprocated later in the same month when the United States Agency for International Development (USAID) hosted a seminar in New Delhi for US power companies.[108] In the following year, AES Transpower, a relatively young US power company that was rapidly internationalizing its businesses, signed an MOU with the government of Odisha to build a 420 MW unit in Ib Valley.[109] By then, SEI's plans to build a megaplant in the same area had been discarded, and the government of Odisha through its electricity board, the OSEB, was building a more modestly sized plant, consisting of two 210 MW units. In October 1993, Pioneer Energy Inc., a consortium of US firms including giant energy companies like Duke, signed an MOU to build a 250 MW plant also in Ib Valley.[110]

Like Odisha, other state governments were also hurriedly signing MOUs with private power companies, again mostly American ones. The most active were Tamil Nadu, Andhra Pradesh, and Karnataka, all states that had huge future power demands due to the growth of the southern information technology hubs in Bangalore, Hyderabad, and Chennai (formerly Madras). For these states, investments in power were meant to support growth in other sectors of the economy; in Odisha, the biggest growth potential lay in the power sector itself. Within Odisha, there was hope that the state could serve as the power source for the economic growth of other states.

Several state-level institutions remained as obstacles to this plan, however. Although the 1991 amendment to the Electricity Act of 1948 opened generation to the private sector, transmission and distribution (T&D) were still in the public sector. From a monopoly, the system was transformed into a monopsony, one in which the sole buyer—the OSEB—was an insolvent actor, saddled with debt, inefficiency, and electricity tariffs that were too low to cover the costs of either generating power in-house or purchasing it from private companies. Even as the heads of foreign companies were busily signing memoranda to build power plants, they were expressing worries about "tardy payment flow" from the SEBs.[111]

Unlike other state governments, Odisha was quick to grant risk-averse private investors a variety of financial and political reassurances. The first reassurance to investors came in the form of a governmental counterguarantee.[112] The government also agreed that private companies could provide power directly to industrial consumers, which would allow private generators to entirely bypass the state-owned distribution utilities. In its negotiations with AES, the government took a further step. After first signing a counterguarantee, the government later signed an additional tripartite agreement with AES and the central government, which guaranteed that if the company's invoice was not paid, the amount would be deducted from Odisha's planned assistance from the central government.[113] The central government itself then signed a counterguarantee. In May 1993, the OSEB signed a thirty-year power purchase agreement (PPA) with AES, the first such agreement in India.[114] Between the PPA and the initial financing arrangements for the deal, it was clear that the government of Odisha was willing to shoulder considerable burdens to ensure the success of what was at the time the largest private investment in the energy sector.[115]

Odisha also took the unusual step of increasing electricity tariffs at a time when most other state governments were reducing or eliminating them.[116] And in late 1993, the Odisha government declared that it would abandon all new generating projects, signaling that it was willing to leave the sector entirely to private firms.[117] At the same time, the private company operating in Kolkata—Calcutta Electricity Supply Corporation—expressed interest in taking over distribution services.

The World Bank in Odisha

It was against this backdrop of ongoing discussions with private firms that the World Bank was negotiating with the Odisha government to fund the Upper Indravati hydroelectric project. The project, which was to serve the needs of electricity generation and irrigation, had fifteen years of prior history. It was first approved by the Odisha government in 1979, and initially received World Bank funding in June 1983. The project suffered several delays throughout the 1980s, in 1991 due to floods, and again in 1992, when a cofferdam built by one of the project's contractors burst. In 1991, the Odisha government agreed at the Bank's behest to reorganize the Upper Indravati project. The Bank suspended its loan for the project in December 1991, in anticipation of the imminent reorganization. Power sector reforms were a precondition to continued funding for the Upper Indravati initially.[118] However, because of delays to the power sector reform project (which required legislative and institutional changes), the Bank's own documentation states that the two were technically unrelated.[119]

In the early 1990s, the World Bank had reevaluated its lending policies in the energy sector, deciding to "lend only to states that agreed to totally unbundle their electricity boards [separating them into discrete generation, distribution, and transmission entities], privatize distribution, and facilitate environmental reform and the private sector's involvement in power generation."[120] From 1993 to 1996, the bank withdrew more than $2 billion in nonperforming loans to entities in the energy sector, and did not make any new energy sector loans.[121] From that point on, all of the Bank's efforts and finances were directed toward restructuring SEBs, a task it started to implement with loans issued between 1996 and 1999, of which Odisha was the first recipient.

Biju Patnaik formally conveyed the Odisha government's commitment to power sector reform in a letter to the Bank in November 1993. The first step of the reform project included improving the financial and operational performance of the OSEB, underwritten by a Bank loan initially granted in 1993, extended twice, and ultimately used in 1994 and 1995. This first loan was also used to establish the reform program, hire consultants, and solicit funds from other bilateral and multilateral agencies, including the Asian Development Bank and the United Kingdom's Department for International Development.

In order to carry out the Bank's restructuring vision, the Odisha legislature had to pass a new law to dismantle the OSEB, create an independent electricity regulator, guarantee that tariffs would be set on commercial terms, and permit private-sector participation in T&D.[122] The law was passed by the assembly in 1995 and ratified in January 1996, after an intervening election in 1995 returned the rival Congress Party to power, bringing back to the chief minister's office J. B. Patnaik, who had occupied it throughout the 1980s. Although the government changed hands from the Janata Dal to the Congress, the new government continued to implement the existing development agenda, including the power sector reforms. One month after taking office in March 1995, the new chief minister issued a formal statement of its power policy on April 20, 1995, committing itself to the previous administration's agenda and outlining a timeline for its completion. In a press conference, J. B. Patnaik expressed the hope that "by the year 2000, Odisha will become the powerhouse of the country."[123]

One important difference between the OSEB and other SEBs was the degree to which the former relied on external sources of generating capacity. The OSEB was heavily reliant on electricity from the NTPC (also discussed in Chapter 2). The few generation assets that the OSEB did own were privatized even before the T&D systems were sold, although the government retained ownership of the hydro sources. When the World Bank changed its lending practices to the energy sector, this had repercussions not just for the SEBs but also for utilities owned by the central government, like the NTPC. The World Bank's requirement that energy utilities operate along commercial principles meant that the NTPC had to better manage its bill collections from its largest customers, the SEBs. One of the strategies the NTPC employed to meet the Bank's requirement was to negotiate with individual SEBs to take over generating plants as a way to eliminate debt. The NTPC reached such an agreement with Odisha in 1995, when the state sold its 460 MW power plant at Talchar to the central utility. The two utilities initially disagreed about the value of the asset, which an outside consultant estimated to be Rs. 3.49 billion ($116 million).[124] Ultimately, the plant was sold for Rs. 3.65 billion ($121.7 million).[125]

The state's other non-hydro-generating capacity was managed by a separate public company, the Odisha Power Generating Corporation (OPGC), which was established in 1984. The OPGC's first plant—a 420 MW plant at Ib Valley—became operational in 1995. It was clear by the early 1990s that the state utility would not have sufficient resources to expand the project further, prompting the government to accept the bid from AES Transpower to build an additional 420 MW plant at the same site. AES's presence in Odisha's generation sector was to play an important role in the course of distribution privatization.

Restructuring and Privatization

In April 1996, after the reform legislation came into the effect, the OSEB was dismantled and the Odisha Electricity Regulatory Commission (OERC) was formed. The OSEB was split into the Grid Corporation of Odisha (GRIDCO) and the Odisha Hydro Power Corporation. The distribution function of the utility was further subdivided into four geographic zones that were to form the basis of future privatization. GRIDCO remained a public company in charge of transmission. Significantly, several other departments within the OSEB were disbanded and not reconstituted elsewhere, like the rural electrification division. Much of the accumulated debt of the electricity system remained with GRIDCO in order that the four distribution companies could show positive net worth prior to privatization.[126]

The formal agreement for the reform program was signed in July 1996 by four parties: the World Bank, the Odisha government, the central Indian government, and GRIDCO. The entire cost of the program was estimated to be $997 million, of which the World Bank's contribution was to be $350 million.[127] In the initial design of the program, privatization of the four distribution zones was to occur sequentially, to benefit from iterative learning.[128] A miniprivatization in 1995 was meant to prove that a private company could increase revenues and operational efficiency. It would also give the state and the private company time to arrive at a mutually agreeable sale price. The government invited bids for a two-year management contract to take over distribution services in most of Dhenkanal and Cuttack Districts and the capital city of Bhubaneswar.[129] Of the four companies that bid, Bombay Suburban Electric Supply (BSES), the private company that at the time operated only in Mumbai, won the contract.[130] The contract was later postponed for one year and transformed into a three-year contract. The greatest challenge for the private company was to lower the T&D losses, which were estimated to be 45 percent; BSES promised to reduce these to 25 percent by the end of three years.[131] A second challenge was to improve billing and collection, which at the time hovered between 75 and 80 percent; BSES pledged to increase this to 95 percent. Less than one year later, in April 1997, the government revoked its contract with BSES on the grounds that the company had failed to honor one of the terms of the deal. According to GRIDCO, BSES's revenue collection from October 1996 to March 1997 had fallen short of its target. For its part, BSES claimed that the failures were the fault of GRIDCO staff, who remained outside of the company's administrative control under the terms of the contract.[132]

There were additional reasons for the contract's termination. When the distribution companies, or DISTCOs, were hived off from the other parts of the erstwhile OSEB, much of the debt was retained with GRIDCO, the company that remained in state hands. This was done to ensure that the distribution segments were attractive for private bidders. From the time that the distribution zones were corporatized, however, they began amassing new liabilities. Some in the government worried that these mounting debts decreased the likelihood of finding buyers.[133] The state government also came to believe that a privatization process stretched out over a long period of time would create needless uncertainties in what was already an uncertain process, the first statewide privatization of a public electricity utility in India.

Prompted by these new concerns, instead of privatizing the distribution utilities one at a time, as had been proposed, the state initiated privatization of all four distribution companies at once in November 1997. The terms of the offer included full protection for current employees of the distribution companies, to avoid opposition from labor unions. Initially, twenty companies expressed interest in the deal by purchasing the government's tender offers.[134] Of these twenty, twelve companies and consortia submitted bids to the government by April 1998 for 51 percent ownership of four distribution companies, known as Western Electric Supply Company (WESCO), Northern Electric Supply Company (NESCO), South Electric Supply Company (SOUTHCO), and Central Electric Supply Company (CESCO).[135] Despite this initial enthusiasm from foreign and Indian companies, in the end, only three firms submitted bids for the four distribution zones.[136]

The government at this point was stipulating that no single company or consortium could take over more than two of the four zones.[137] By April 1999, however, the government was forced to change this policy due to the paucity of reasonable offers. Ultimately, BSES assumed 51 percent ownership of three distribution companies, in the west, south, and north. All are regions where there were few residential and agricultural consumers. GRIDCO would continue to own 39 percent equity in the firms, and an employees' trust would hold the remaining 10 percent.[138] For the fourth zone, in central Odisha, the government received an offer from a consortium of Tata Electric Companies and UK-based Viridian Company. The consortium asked for the government to issue a six-month letter of credit, which was meant to safeguard the company from an additional burden placed on this zone. In an earlier agreement, revenues from this zone were guaranteed to an escrow account to pay for power generated from the OPGC, which was the result of the earlier privatization of this generating unit.[139] Since the condition placed by the Tata-led consortium was deemed unacceptable to the government, the only viable remaining buyer for the central zone was AES itself. An earlier ban that prevented generating companies in the state from bidding for the distribution zones was lifted, clearing the way for AES to take over CESCO, the central distribution company, as of September 1999.[140]

GRIDCO remained as a state-owned company in control of transmission functions in the state, which entailed buying power from generating plants and moving it along high-tension wires to the distribution companies' transformer sites. The state remained, therefore, as a middleman between four private distribution companies on one side, and generating companies like private-sector AES and the central government's NTPC on the other. From the start, all four of the new private distributors expressed dissatisfaction with the terms of the sale and the state's tariff policy, which was established by the Odisha Electricity Regulatory Commission. On each occasion of a tariff revision, the distributors asked for a much more substantial tariff increase than the OERC ultimately awarded. Furthermore, the private companies claimed that the system losses were much higher than what the government had told them prior to privatization, on the basis of which they had valued and bid for assets. According to the project estimates, transmission and distribution losses averaged 39.5 percent in 1996–97. The reforms were meant to reduce these to 23 percent by 2002–3.[141] But almost immediately after assuming control of distribution, the private companies upwardly revised the T&D losses (see Table 4.3).

While three of the private entities owned by BSES continued to operate, AES, which owned CESCO, eventually abandoned the distribution project entirely in 2001. As a distributing company, AES refused to pay for the

TABLE 4.3
Post-privatization estimates of T&D losses (%)

	CESCO	*NESCO*	*WESCO*	*SOUTHCO*	*GRIDCO's estimate of total system losses*
1996–1997	52.9	44.4	42.1	45.1	49.5
1997–1998	47.9	42.1	38.4	35.2	49.2
1998–1999	48.6	44.6	44.6	43.7	48.6
1999–2000	44.8	43.4	44.2	41.9	43.8
2000–2001	44.9	44.4	43.9	42.5	43.4

Source: "Report of the Committee on Power Sector Reform of Orissa," sec. 4.21.

power that it purchased from GRIDCO, which in turn claimed that its loss of revenues meant that it could no longer pay for the power that it purchased from AES's generating plants.

There were numerous reasons that this part of the privatization program failed. The terms of the contract governing the sale were unclear, leaving loopholes for malfeasance on the part of the government as well as the private owners. And according to an audit by India's comptroller and auditor general, the government showed AES undue favor in the terms of the deal, suggesting the considerable lengths that the government was willing to go to in order to find sufficient private buyers.[142] AES cited myriad reasons for its dissatisfaction, including "deep-rooted flaws in the regulatory, policy and administrative environment."[143] The company also complained of recalcitrant employees, who were protected under the terms of the contract, and government and bureaucratic delays in releasing World Bank loan funds meant to be used for making capital investments in the distribution system. The company also protested that some of its most delinquent customers were state enterprises and municipal bodies.[144] While the privatization program had support at the very top levels of Odisha's governing class, those in the middle ranks of the government and bureaucracy were not as supportive. For its part, the government, GRIDCO, and the OERC complained that from the start, AES (and also BSES) failed to improve operational efficiency and invest in capital improvements.[145]

Conclusion: The Winners and Losers of Privatization

The Odisha government's power sector reforms earned the state high marks as an agile reformer, and were useful as a way of spurring private investment to the state. At the Indo-US business summit held in Calcutta in 1997, the

US secretary of commerce, William Daley, predicted that Odisha's capital would be Southeast Asia's most important commercial city in the next century.[146] Private capital was equally hopeful, as the size and quantum of proposed investments described earlier attests.

The primary "losers" of the restructuring and privatization programs were the communities and households that had never been connected to the state's electricity grid, which are a substantial number in Odisha. These were disproportionately located in the inland districts (although even the most electrified districts in Odisha still had very low rates of household electrification compared to prosperous districts in other states) and composed of SC and ST communities.[147] As mentioned earlier, the rural electrification wing of the erstwhile OSEB was disbanded and not reconstituted in any other state government ministry. The government may have hoped that private distributors would carry on with rural electrification, but did not provide any institutional or financial incentives to encourage this. In reality, rural electrification came to an almost complete halt following privatization and began again only in 2005, under a new set of central government funding initiatives to provide public goods to rural India.[148]

Others who were adversely affected by privatization were urban domestic and commercial consumers, whose tariffs rose steadily throughout the 1990s. The government raised tariffs as early as 1992 to signal its commitment to transforming the sector. From 1992 to 2001, the average tariff increased 267 percent, although the increases were not the same for all categories of consumers. Initially, the tariff increases were concentrated on the industrial and commercial consumers, with the guarantee that the state would reduce the duration of forced power outages from ninety minutes each day to thirty minutes.[149] But from 1996–97 to 2000–2001, the largest increases were shouldered by domestic, commercial, agricultural, and small industrial consumers. Tariffs for large industries remained fairly stable over this period.[150] It is likely that for industrial and commercial consumers, particularly the medium to large ones, the trade-off between better-quality supply and slightly higher prices was an acceptable one. For the 50 percent of households subsisting below the poverty line in Odisha who had access to electricity, however, even a modest increase in tariffs would have constituted a deep financial strain.

The project proposal agreed to by the World Bank and the government in 1996 states that the most "important beneficiaries" would be the industrial consumers:

> Industrial consumers in particular will benefit from the improved quantity, quality and reliability of their power supply, since they account for the majority of consumption and bear the brunt of load shedding [planned power shutdowns] and other failures of supply. The Orissa government expects the state's improved power supply outlook to attract additional industrial investment and encourage existing industries to expand their production facilities. This expectation is fully in line with national surveys where industrialists consistently rate power supply as one of their most critical constraints.[151]

In the lengthy, 222-page document, which includes details of the loan agreement, the reform program, and numerous appendixes on various aspects of Odisha's power sector, there is very little discussion of the low rates of electrification that characterize much of Odisha. A later section states that the program is expected to only "*indirectly* help the poor by freeing up state funding . . . for higher priority use in the social sectors."[152] The privatization program, then, was never expected to benefit either rural consumers or underserved communities in the immediate term, or directly.

On the face of it, utility privatization in a state with such an underdeveloped grid seems to make little sense. But it was precisely the underdeveloped nature of the sector that made privatization a politically permissible strategy. There were few protests from either consumer groups or movement activists, and even fewer from the kinds of farmers' organizations that would play such an important role in bringing privatization to a halt in Andhra Pradesh. It helped too that since this was India's first major utility privatization, there was little awareness of the repercussions for ordinary citizens. Finally, the utility privatization makes sense when read alongside the state's longer politico-economic history, which shows a string of repeated attempts to take advantage of the state's mineral resources to initiate large-scale industrialization.

5

Social Movements and Electric Populism in Andhra Pradesh

If the state in Odisha was a model of a regional political regime committed to rapid heavy industrialization and willing to devote development resources to achieve that end, then the development policy regime put in place in Maharashtra from the 1960s to the 1980s emphasized the need to balance what was already a robust industrial sector with an equally vibrant rural and agro-industrial economy. The state in Maharashtra directed electrification resources toward increasing agrarian productivity and then promoted a decentralized agro-industrial economy focused around sugar manufacture in the western districts of the state. The pace and direction of electrification in each case functioned not as a technocratic exercise but rather as part and parcel of an emerging development strategy. While Odisha continued to be dominated by a largely urban-based, educated, and entrepreneurial class, the state in Maharashtra was commanded by a coterie of rural leaders with strong ties to agriculture and agro-industry who had themselves benefited from the levers of state development patronage, the sine qua non of which were the cooperative credit and processing institutions. The extent to which rural classes had penetrated the provincial political establishment helps make sense of whether and how much development spending and electricity spread through the countryside in those two states.

Andhra Pradesh, however, tells a different kind of story about how the political strength of the countryside impacted the state's development commitments and priorities. In Maharashtra, the political rise of the largely rural Maratha caste cluster was coterminous with—and even to some extent preceded—state formation in 1960. In Andhra Pradesh, by contrast, rural development came to the fore of the state's policy agenda once a successful farmers' movement emerged in the 1970s. Initially operating outside of the state, groups that had felt marginalized by successive Congress-dominated governments agitated for a number of inputs and price supports, chief among which was subsidized and plentiful electricity to operate irrigation pumps. In the 1980s, these disparate social groups found new political representation in a regional party spearheaded by the charismatic movie star turned politician, N. T. Rama Rao, or NTR.

In the decades after Andhra Pradesh was formed, the state and the utility took a measured approach to rural electrification, exhibiting neither the neglect of the rural sector found in Odisha nor the enthusiasm for energizing irrigation pumps of Maharashtra's development regime. Andhra Pradesh started off with much higher rates of village electrification thanks to colonial-era commitments to rural electrification when the coastal districts of the state were part of colonial Madras. Although village electrification was more extensive in Andhra Pradesh than in other parts of the country, the tariffs charged for rural customers were calibrated to the costs of supplying electricity to them, which meant that electricity tariffs for farmers were substantially higher than those for industrial consumers. Furthermore, in the first few decades after the Andhra Pradesh State Electricity Board (APSEB) was created, there were regional differences in tariffs; farmers in the coastal districts of Andhra Pradesh paid lower prices, reflecting the fact that the costs of supplying these regions was lower.

The cost and availability of rural electricity improved in the aftermath of a series of farmers' movements throughout the 1970s that demanded lower rates for electricity. These farmers' movements comprised one of the social processes that contributed to making Andhra Pradesh's electoral politics a competitive space in the 1980s. The charismatic NTR, whose Telugu Desam Party (TDP) finally ended the Congress Party's hegemony in the state, capitalized on the discontent of the countryside, pledging lower electricity rates as an important part of a large bundle of populist promises. The result was a marked shift in the state's development priorities during the late 1970s and early 1980s. Andhra Pradesh's earliest plan documents stress the importance of electricity for industrial development, with little emphasis on how electricity transforms both rural productivity and rural livelihoods. It was only after the farmers' movements in the state demanded electricity as a vital agrarian input that the state's development machinery responded with greater attention to rural consumption.

The history and politics of electrification from the 1960s to the 1980s leave a strong imprint on the processes of energy sector liberalization in the 1990s in Andhra Pradesh. As Chapter 3 argues, Maharashtra was expected to take an early lead in implementing electricity distribution privatization because the state had already achieved rural electrification, had leaders who were open to other kinds of market reforms, and could have easily attracted private investors. The nature of the Maharashtra state's ties to rural constituencies, however, meant that the government shied away from privatization because it would have threatened to dismantle a long-standing system of state subsidies to politically influential large farmers and agro-industrialists.

Like Maharashtra, Andhra Pradesh too was expected to be a leader in economic liberalization, including in the electricity sector. The reason for this belief was bound up in the figure of N. Chandrababu Naidu, who was the state's chief minister from 1995 to 2004, the entire period when distribution privatization was being both advocated (by the central government and the World Bank) and enacted (by Odisha and Delhi). Hailed a "CEO" chief minister by the Indian and global business press, as well as by international financial institutions, Naidu professed a strong ideological commitment to economic liberalization. He seemed to evince an iron "political will" to liberalize, which was considered the necessary factor in many accounts of economic reform.

Naidu's Andhra Pradesh was one of three states (along with Odisha and Haryana) to embark on the process of electricity sector restructuring that had been advocated by the World Bank starting in the mid-1990s. The process, as discussed in previous chapters, had two primary steps. The first involved separating the state's vertically integrated public utility into separate transmission, distribution, and generation companies. The distribution business was then to be further subdivided by geographic zone. According to the restructuring plan followed by Odisha, transmission was intended to remain under state ownership, generation would become a mixed sector in which new private companies operated alongside the older state-owned plants, and distribution would be privatized with the ultimate goal of creating a competitive distribution environment. Naidu's government was able to initiate the process of restructuring much more quickly than most state governments by passing the Andhra Pradesh Electricity Reform Act in 1998. However, the state's rural population consistently resisted Naidu's efforts to prepare the ground for privatization. Rural voters for whom electricity subsidies were gradually being decreased were spared the full brunt of the tariff increases that affected other categories of consumers, even though both Naidu and the World Bank funders viewed rural electricity consumption as the primary cause for the utility's operational and fiscal difficulties. In response to the very minimal tariff increases that affected the farm sector, Naidu's political opponents depicted his government as antipoor and antifarmer. After serving two terms, the CEO chief minister lost in 2004. A significant feature of the Congress Party's election platform was a promise to restore free power for farmers, and implementing this promise was the first act of Andhra Pradesh's new chief minister, YSR Reddy. This, together with the termination of World Bank funding, signaled an end to Andhra Pradesh's experiment with electricity privatization.

Naidu's attempts to restructure and privatize APSEB speak to his ideological commitment to reform (necessary to stimulate industrialization and "development," as the chief minister himself often put it). But these commitments ran up against the political pressures to maintain populist subsidies that benefited the state's rural communities. Just as in Maharashtra, electricity had become an important input in agricultural production, particularly in the rich agricultural districts of coastal Andhra Pradesh. Again, as in Western Maharashtra, agricultural production in these districts had been stimulated in an earlier period by British colonial spending on canal irrigation; in the postcolonial era, these districts were also the prime beneficiaries of Green Revolution growth, particularly through water-intensive rice cropping. The result of increased spending on rural electrification in the 1970s was that a steadily increasing percentage of the state's electricity supplies were allotted to agricultural consumption. And agriculture in Andhra Pradesh relative to other states is particularly reliant on tube well irrigation via electrified pumps. The large consumption of electricity by the farm sector was the critical factor affecting the state's energy policy; it was blamed for the fiscal crisis of the public utility and thus became the most urgent justification for reform. And it was also the biggest obstacle to the implementation of the restructuring and privatization programs that were being supported by the World Bank's lending program. Naidu's power sector reforms also became the most visible symbol of his government's economic agenda and served as a lightning rod for political opposition, bringing parties of all persuasions together to launch massive protests and rallies. According to media sources, Naidu's unpopular power reform policies stimulated the longest agitation in the state centered on an economic issue.

State Formation in Andhra Pradesh

Like most other states in India, Andhra Pradesh is an amalgam of regions, each with distinct ecology, social structure, and political history, but united by a common linguistic identity—in this case, the language Telugu. The three different regions of the state—Rayalaseema, Telangana, and coastal Andhra—were part of different political formations during the colonial period. Telangana was the seat of the Nizam's Hyderabad, and contains the state's capital city, Hyderabad. At the time of state formation, most of the industrial activity was located in and just outside of Hyderabad, but the remaining districts of Telangana consisted of rain-fed agriculture. The Rayalaseema region, which includes the southwestern districts of the state, was ceded by the Nizam to the British and was administered as part of Madras Presidency. Like Rayalaseema, coastal Andhra Pradesh (in the eastern part of the state) also became part of colonial Madras Province. The districts of coastal Andhra during the colonial period benefited from extensive irrigation investments in the late nineteenth and early twentieth centuries, and were among the most productive agrarian areas in south India. As a consequence, these districts were well poised to take advantage of the Green Revolution policies of the 1960s, which further exacerbated the inequalities between agriculture in coastal Andhra and the other two regions. Rayalaseema, which has a distinct agricultural climate from the coastal districts, neither benefited from irrigation development nor had natural resource endowments that could be exploited.

The geographic contours of Andhra Pradesh were assembled in two stages. Following an impassioned linguistic movement demanding a separate state for Telugu speakers, the Telugu districts were carved out of Madras Province into a separate Andhra State in 1953. The new state's capital city, Kurnool, was located in the coastal district of the same name. Three years later, after the States Reorganisation Commission submitted its report, the remaining Telugu-speaking districts that were part of the Nizam's Hyderabad (also called Telangana) were included as well, and the capital city was shifted to Hyderabad (see Map 5.1).

The electoral history of the state during a sixty-year period has two distinct phases, each lasting roughly half of that time. For the first half, lasting from the time of state formation in 1956 until elections in 1983, the Congress had immense political success in the state. The party prevailed in Andhra Pradesh even through the 1967 elections, when it lost at the state level in many other states, as well as in 1977, when disaffection with Indira Gandhi's

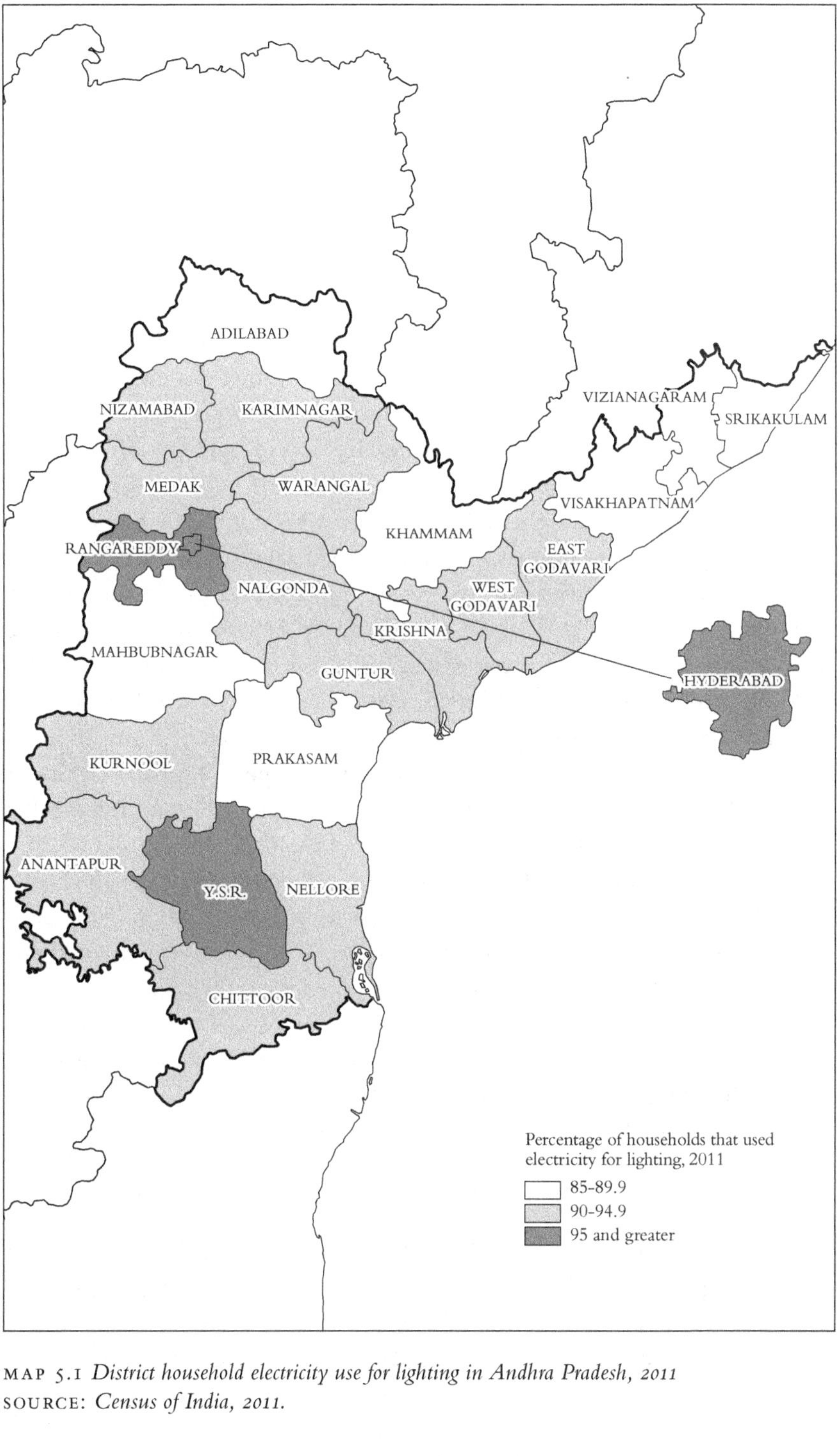

MAP 5.1 *District household electricity use for lighting in Andhra Pradesh, 2011*
SOURCE: *Census of India, 2011.*

"Emergency" ruined the Congress's political fortunes in most other parts of the country. Although the leftist parties had made deep inroads in the districts of modern Andhra Pradesh from the late colonial period into the 1950s, Congress was able to undercut their support base by first conceding the demand for a linguistic province, and later by enacting land reforms, both of which had been important mobilizing initiatives of the Communist parties.[1]

After 1983, the political landscape shifted decisively. The emergence of a second political force in the regional TDP marked the beginning of a highly competitive political field. Underlying these changes to the electoral field were social changes, including the rise of new movements articulating a new set of demands. The most consequential for the state's political economy of distribution and development were the farmers' movements of the 1970s. NTR picked up on the political ferment represented by these vocal and organized farmers groups from across the state, which demanded state support for higher prices and subsidies for vital agrarian inputs, of which rural electrification and electricity tariffs were a central concern. In addition to these concerns, NTR also capitalized on the dissatisfaction of one of Andhra Pradesh's two dominant peasant castes, the Kammas, a community that had felt sidelined, if not exactly excluded, from the field of political power.

Under NTR, the TDP's electoral campaign consisted of a host of populist promises that were designed to satisfy both the poorer urban constituencies (such as the promise of rice for two rupees per kilo) and the wealthy among the rural populace (such as subsidized electricity for farmers).[2] After winning office, NTR's highly centralized regime tried to satisfy the party's main bases of support. Electricity tariffs, which had already begun to decline for farmers relative to other categories of consumers (chiefly industrial and commercial) in the late 1970s, dropped even more precipitously after the TDP came to power, with the expansion of free electricity to farmers. These promises had consequences for the overall stability of the electricity sector, the public utility's financial viability, and the state's ecology, which experienced rapidly declining groundwater tables.

The Electricity Sector and APSEB

APSEB was formed in 1959. Initially, the new government of Andhra Pradesh kept the functions of electricity generation, exploring new generation projects, and maintaining the largest of the transmission lines within the Electricity Department of the government, and only transferred the functions of electricity distribution and supply of power to consumers to the newly created APSEB. By 1963, all electricity functions were transferred to the board, although the government continued to maintain control over two hydroelectric projects that were shared with other states: the hydroelectric stations at Machkund (located in and shared with Odisha), and Tungabhadra (located in and shared with Karnataka). The board was set up with four members, two of whom were Indian Administration Service (IAS) officers and two of whom were electricity engineers. Functionally, the activities of the Electricity Department and the board were thoroughly interconnected, as the board shared the administrative personnel of the Electricity Department.[3]

As in Maharashtra, at the time that the state was formed, all of the available electricity supplies were concentrated in Andhra Pradesh's cities and towns, and in some cases, in the immediately surrounding rural areas. But unlike in Maharashtra, APSEB rapidly nationalized all of the existing private electric utilities in the first three years after it was formed. By December 1961, when the municipal utilities in Vijayawada and Masulipatnam were taken over by APSEB, Andhra Pradesh had an exclusively state-owned utility system.[4]

From the 1960s onward, the expansion of the state's electricity generation, distribution, and transmission was a centerpiece of Andhra Pradesh's development effort. This is evident right from the early plan periods. In the *First Five-Year Plan* period, during which the districts of Andhra were part of Madras State, 39 percent of the state's plan was devoted to power. In the *Second* and *Third Five-Year Plan* periods, the plan allocations for power were lower at 20 and 27 percent, but still larger than the budgetary allocation for any other sector. To give a sense of the importance of electricity to the planning exercise, consider that during the *Third Five-Year Plan* period, a total of only 15 percent of the budget was devoted to the entire social services sector combined, and only 5 percent for road transport in the state. The Planning Department in Hyderabad, in a retrospective analysis of the first three plans, acknowledged the high "social cost" of placing so much emphasis on power, but justified its plan allocations by noting that they are in line with the "interests of national priorities."[5] The heavy emphasis on power was even more pronounced in subsequent five-year and annual plans. From 1974 to 1980, a period that includes the *Fifth Five-Year Plan* period, out of a total plan of 1,892.46 crores, 689.01 crores were devoted to the power sector, which represents 36.4 percent of the total plan spending and three times the amount spent on social and community services combined.[6]

Despite this heavy spending on the electricity sector, though, APSEB took a very measured approach to rural electrification. From the 1960s to the early 1970s, the kind of political interference and pressure to electrify the countryside that plagued the finances and the technical stability of Maharashtra's grid were absent in Andhra. As the next section relates, the board's implementation was more measured compared to the rapid progress of rural electrification in other states. Instead, during the 1960s, Andhra Pradesh was like Odisha in trying to entice new industrial investors with the promise of concessional electricity tariffs.[7] Preferential policies for agrarian consumers began only in the late 1970s, increasing first in the wake of farmers' movements demanding a greater distribution of the state's development resources, and then continuing through the 1980s.

Lighting the Village and Energizing the Pumps

If the largest consumer of water in Maharashtra was the water-hungry sugarcane crop, the culprit in Andhra Pradesh was rice, or paddy. Commercial agriculture, including paddy, had expanded after the first investments in canal irrigation in the late nineteenth century, and became even more extensive after independence, particularly during the Green Revolution. While farmers in Punjab came to dominate wheat cultivation and reaped the benefit of the central government's public wheat procurement program, the same was true of Andhra Pradesh's rice farmers, particularly in the coastal districts. From the mid-1950s to a high point in 1998–99, the cropped area in the state under rice cultivation increased by close to 50 percent.[8] In roughly the same period, the share of coarse cereals, many of which are drought-resistant and more suitable to Andhra Pradesh's ecology, decreased from 40 percent in the 1960s to 11 percent in the late 1990s, a more rapid decline than in India as a whole.[9] The politics that led to these cropping outcomes in Andhra Pradesh are part of a general pattern that mirrors some of the trends seen in Maharashtra, where politically influential agricultural communities were able to channel subsidized resources to input-intensive cash crops.

Unlike in Maharashtra, however, rural electrification expenditures and subsidized electricity for farmers emerged not at the moment of state formation, but only fifteen years later, after vibrant farmers' movements demanded a greater share of the state's development expenditures for agriculture. In Maharashtra, from the outset of state formation in 1960, the dominant peasant Maratha caste used its political representation in the government and privileged berths in cooperative bureaucracies to widen the channels of state investments in agriculture, particularly evident in the state's robust rural electrification program. In Andhra Pradesh, by contrast, the state took a more restrained approach to rural electrification, energizing pump sets and extending the grid to rural locales on a steady but not rapid pace throughout the 1960s and early 1970s. In addition, the electricity rates charged to Andhra Pradesh's farmers were on average higher than those charged to industrialists in the state. It was only after the mid-1970s that both the rates of rural electrification and the electricity tariffs for farmers started to skew in favor of farming communities.

A couple of measures are useful in judging APSEB's development priorities, as well as in determining decisive points of departure from existing patterns of spending. The rates at which villages are electrified and irrigation pumps are energized give a sense of how, and how quickly, electric utilities tackled the overall problem of electrifying the countryside. In Maharashtra, the rates of rural electrification were high throughout the 1960s, ratcheted upward by local politicians who used political pressure to get their constituencies electrified, as well as by subsidized credit mechanisms that facilitated the spread of electrification for the purposes of irrigation. In Andhra, by contrast, during the 1960s, the board at several points lamented that the pace of rural electrification was unable to keep up with the demands from farmers. In its annual reports, presented each year in the legislative assembly, the board's administrators noted that either the state or the central government must step in to provide additional financing for rural electrification. In at least the first decade of the board's operation, there is no evidence that cross-subsidies (lowering the costs of one set of consumers by raising the costs for another) were used to finance rural electrification. In the first annual report of the board, published in 1960 after the first full year of the utility's operation, the tariffs for the largest industrial consumers (those who are served by high-voltage wires) ranged from 4.7 to 2.6 paise per kWh of electricity consumed, with lower marginal charges for the largest volumes of consumption. For agricultural consumers, all of whom relied on low-voltage supplies, tariffs ranged from 9 to 6 paise per unit of electricity consumed, again with lower charges levied for farmers who consumed the greatest quantities.[10] The board's tariff structure reflected the costs of supplying different categories of consumers. Industrial consumers—who were generally concentrated in urban or periurban locales and consumed very large quantities of electricity—were much less expensive to serve than the more numerous but less power-intensive agricultural consumers. Laying transmission and distribution (T&D) lines to far-flung rural locales was a costly operation, one that was at least partially reflected in the tariffs that the board charged to farmers. Throughout the 1960s, the annual reports of APSEB acknowledge that the pace of rural electrification is very slow in the state. About the high cost of electrification and the social function it serves, one APSEB report states:

> An analysis of the financial position of the State Electricity Boards reveals that despite the relatively high priority assigned in most states to rural electrification, inadequacy of funds has been hampering their efforts. The immediate revenue returns are inadequate from the commercial point of view. But considering the value of the economic and social benefits which are bound to accrue to the community from investment on rural electrification, it is important that this programme is given a high priority in planning for economic and social development. The fact that from month to month applications are pouring in for energisation of pumpsets is a clear indication of the farmers' awareness of the importance of power for lifting water for irrigation and their genuine eagerness to do intensive cultivation with assured water supply, to produce more and to live better through self-help.[11]

At a time when the utility in Maharashtra complained that politicians were putting undue pressures on the board to electrify pump sets and extend the rural grid, APSEB complained that the state was underfunding rural electrification. The passage also demonstrates that farmers in Andhra Pradesh were eager to embrace this new technology. In proposing a model for financing rural electrification, the annual report's authors note that public utilities everywhere have required political intervention to provide subventions and other measures of financial support to what is almost everywhere a losing proposition in the initial years: "Even in countries like the United States of America from the beginning, it was appreciated that rural electrification would require Government's participation. Unless either long-term interest-free loans are given for rural electrification or part of the cost is met by the Centre as subsidy, rural electrification is not likely to be self-supporting for some more years to come."

The rate of rural electrification in Andhra Pradesh increased after 1969, when the central government formed the Rural Electrification Corporation (REC). The REC provided an important new means to finance rural electrification that did not take funds away from the board's other priorities of augmenting generation supplies and encouraging industrialization by providing connections to industrial consumers. But despite the evident fillip provided by REC funds for rural electrification, during its first twenty years of operation, the utility's overall pattern of allocation to different consumers

TABLE 5.1

Percentage electricity sales in Andhra Pradesh by category of consumer, 1960–1961 to 1975–1976

	1960–1961	*1965–1966*	*1970–1971*	*1975–1976*
Domestic	10.8	8.1	8.2	9.5
Commercial	7.5	8.2	5.8	4.9
Public lighting	2.3	1.9	1.6	1.1
Industrial—low-tension	8.4	7.8	7.3	8.2
Industrial—high-tension	56.3	56.1	58.7	55.6
Agricultural	9.4	17.4	18.0	20.7
Misc.	5.2	.6	.4	.1

Source: Andhra Pradesh State Electricity Board, *Administration Report*, 1970–71 and 1975–76.

varied little. Industrial consumption outpaced that of other constituencies, and industrial consumers continued to represent the largest category of consumers (see Table 5.1).

For farm consumption, the most significant change occurred in 1982, when the utility authorized a flat-rate tariff fee for farmers based on the horsepower of their irrigation pumps (see Table 5.2).[12] The combination of larger numbers of pumps and lower tariffs for agricultural consumers led to a greater proportion of all electricity consumption in the state going toward agriculture from the 1980s onward. As in other states, the flat-rate system also had negative consequences for the utility's bookkeeping by allowing a huge quantum of electricity use to be estimated rather than accurately measured. Over time, APSEB used the category of agricultural consumption to account for electricity losses that were the result of aging and inadequate T&D equipment, and theft and corruption on the part of utility staff.[13]

By the late 1970s and early 1980s, when rates and access for farmers improved, there were new forms of financing from the central government (through the REC) and a more mature electricity grid with much larger generating capacity than when APSEB was first formed. These technological and institutional improvements were important drivers of the expanding access to rural electricity. Another important part of the story of rural electrification in Andhra Pradesh, though—and, in particular, in explaining changes in how the state apportioned development resources among urban and rural, industrial and agrarian consumers—was the emergence of farmers' movements in the late 1970s and early 1980s.

TABLE 5.2
Electricity tariffs for low-tension agricultural consumption, 1977–1999

	Metered tariff		
Date	*Per-unit rate (paise/kWh)*	*Additional fixed charges (Rs./hp)*	*Flat-rate tariff*
1975–1977	16 paise/unit	3 Rs./hp	
1977–1982	16 paise/unit	2 Rs./hp	
1982–1989	16 paise/unit	2 Rs./hp	Rs. 50/hp
1989	41 paise/unit	6 Rs./hp	Rs. 50/hp
1990–1992			Up to 5 hp—no charge 5 hp–10 hp—Rs. 100/hp/annum
1992			Up to 5 hp—Rs. 100/hp/annum 5–10 hp—Rs. 250/hp/annum; Above 10 hp—Rs. 400/hp/annum
December 1992–1995			Up to 5 hp—Rs. 75/hp/annum 5–10 hp—Rs. 250/hp/annum Above 10 hp—Rs. 400/hp/annum
1996–1999			Up to 3 hp—Rs. 150/hp/annum 3–5 hp—Rs. 250/hp/annum 5–10 hp—Rs. 350/hp/annum Above 10 hp—Rs. 400/hp/annum

Source: Andhra Pradesh State Electricity Board, *Annual Report*, various years.

Farmers' Movements and Farmers' Power

As described previously, the farmers' movements in Andhra were part of the larger process of agrarian politicization. The nature and content of the movements differed according to time and place, and farmers' demands for state support took varied forms. For example, the Bharatiya Kisan Union's organizing in Punjab and the Shetkari Sanghatana's protests in Maharashtra during the 1980s largely focused on the issue of procurement prices, demanding that the state increase the prices at which it purchased food supplies for the public distribution system. In Uttar Pradesh, the Bharatiya Kisan Union organized one of its most widely publicized campaigns on the issue of electricity in 1987. As in other parts of the country, farming in western Uttar Pradesh is heavily dependent on groundwater. When the state government raised the electricity rates for farmers relying on electrified pumps by 50 percent, the union responded by organizing a *dharna*, or protest, that reportedly brought 80,000 farmers to Shimla, about 100 km from New Delhi.[14]

The demands from farmers' groups in Andhra Pradesh were equally varied. In January 1976, a regional farmers' organization based in Andhra

Pradesh, the Raitu Sangham, organized a series of demonstrations throughout the state to demand increased procurement prices, lower agricultural tax rates, and greater access to and lower prices for electricity.[15] More than 20,000 farmers participated in the meetings, which were held in seventy-seven towns across nineteen districts in the state. In June and August 1977, conferences were organized first in Ananthapur and then in the state capital, Hyderabad, devoted to the issue of electricity. Following the latter meeting, Raitu Sangham formed a committee to negotiate with the state government on the subject of electricity prices and access. The Bharatiya Kisan Union (BKU) established an outpost in Andhra Pradesh in 1979. The Andhra BKU staged a rally in front of the state assembly in Hyderabad, again to take up the issue of electricity prices, as well as agricultural debt, crop insurance, and drought relief. One scholar of the Andhra farmers' mobilization suggests that this 1979 rally was crucial in convincing the state government to impose a flat-rate pricing system for farmers' electricity consumption.[16]

The beneficiaries of the greater state support for rural electricity consumption were not evenly spread. Even during the 1960s, village electrification proceeded more quickly in the regions of coastal Andhra and Rayalaseema than in Telangana.[17] Coupled to the geographical bias in favor of coastal Andhra, farm subsidies disproportionately benefited wealthier farmers, a tendency that was reinforced by the flat-rate system, which favored farmers who could afford more-powerful pumps.

Subsidizing electricity use by the farm sector had an effect on the state's overall economic development, particularly in the balance of support for agriculture versus industry. Increasingly throughout the 1970s and 1980s, the supplies of electricity to industrial consumers were curtailed to allow greater allocation to farmers. Unlike in Maharashtra, where at least some industrial consumers were shielded by the barriers that surrounded the privatized Mumbai zones, industrial consumers in Andhra Pradesh relied on the same scarce resources as farmers. In the near term, the effect was a zero-sum game between competing classes of consumers. The cuts for industrial consumers began in the 1970s, as the rural electrification program of the 1960s and 1970s was resulting in a greater demand for electricity and putting pressure on the overall capacity of the grid. In April 1975, the government increased a power cut that was already in place on industrial consumers by an additional 10–20 percent. APSEB also introduced a restrictive weekly quota system. Industrial consumers who exceeded their weekly ration would lose their power connection.[18] According to the state utility's annual reports, for the twenty-three years from 1977 to 2000, there were only nine years in which electricity consumers were not faced with power restrictions.[19] During the 1970s and until the mid-1980s, all consumers shared equally in the cuts, which were fixed at anywhere from 20 to 30 percent. From the mid-1980s through the rest of that decade, there was a noticeable skew in the way that power cuts were distributed among consumers. Those consumers who required the greatest amounts of energy (generally industry and manufacturing) faced restrictions of anywhere from 30 to 60 percent. For consumers who required smaller amounts of power, which would include farmers and domestic users, APSEB imposed a much more modest restriction of anywhere from 15 to 30 percent. During the 1990s, farmers and domestic consumers were entirely protected from the load restrictions that continued to affect large industrial consumers.

This continued to be the scenario as the reform decade of the 1990s unfolded. In what was an annual ritual, in January 1991, APSEB cut electricity supplies to high-tension consumers (the largest consumers of electricity) by 30 percent.[20] January is the height of the sowing season and farmers are therefore urgently in need of irrigation. In addition to the increased needs of farmers, shortages in coal supplies exacerbated the state utility's problems. The power cuts were even higher in 1996, when power to some industries was restricted by between 40 and 60 percent. The reliability of power supply to industry improved only after 1998, when the power cuts on industrial consumers were lifted as part of the market reform program. For a state trying to improve the financial health of the utility in preparation for privatization, it made the most sense to supply as much power as possible to the utility's most lucrative consumers—those who paid the highest tariffs and consumed the greatest quantities.

As suggested by the following narrative of Naidu's attempts to restructure and privatize electricity distribution, his government was ultimately unable to make the necessary changes that would have led to greater quantities and better quality of power to attract new investments in the state. To get around this, his government embraced several policies that favored industrial and commercial establishments by stealth rather than through a frontal attack. Naidu's government also embraced the model of special economic zones that allowed firms to bypass the restrictions that applied to the state at large. For example, in his bid to turn Hyderabad into "Cyberabad," Naidu gave significant concessions on land and infrastructure, including power, to information technology firms.[21] In addition, the government adopted a lenient approach to allowing industrial units to establish captive power plants and sell excess power back to the state grid.[22] The permissive regime around captive power generation has been likened to a de facto privatization of the system.[23] These measures went some way toward attracting the kinds of private capital flows that Naidu's government believed were critical for the state's future development. Changing policies for the entire state, however, was a much more entrenched problem.

Andhra Pradesh's Power Sector Reforms in the 1990s

Andhra Pradesh was at the leading edge of the World Bank's new focus on state-level lending that began in the mid-1990s. It is a poor, populous state, but it also had at its helm a politician in Naidu who set out to reorient the state's policy agenda toward development. By "development," Naidu envisioned engaging the latest technologies in both governance and economic activity, gradually easing the government out of productive spheres and into a more regulatory capacity, and allowing private capital to finance future growth. Andhra Pradesh's power sector reforms, assisted by the World Bank's loan for a power-restructuring program, then, were just the largest part of a broad array of reforms programs fueled by Bank lending. From mid-1998, when the Andhra Pradesh Economic Restructuring Project was initiated, through 2004, when Naidu lost the chief ministership, the volume of Bank lending to Andhra Pradesh totaled nearly US$1.6 billion. According to both Naidu and the World Bank, among all of the various reforms programs, the power sector reforms were the most important.[24]

A World Bank document from 1997 first defined the parameters for the myriad reforms programs that would follow. After a long section in the report dealing with the state's finances, the bulk of the substantive part of the document itemizes six critical sectors that required restructuring, including power, irrigation, roads, ports, education, and health. In the section on the power sector, the report's authors identify a critical power shortage as the main challenge for the sector, and further argue that the power crisis can only be averted by private investment flows. These, however, are unlikely to materialize in the current scenario of massively subsidized tariffs and inefficient financial and physical operations. In order to attract private capital, the report prescribed that the state must carry out five main procedures: "(a) adjust agricultural tariffs to reflect costs; (b) privatize distribution progressively; (c) create an independent and transparent regulatory system; (d) corporatize power utilities and manage them as commercial entities; and (e) enact reform legislation to enable implementation of the above reforms."[25] With respect to tariffs, the document further states that in order to improve the financial position of the distribution companies prior to privatization, tariffs for agricultural consumption must be increased to at least 50 paise/kWh in the short term and progressively increased so that the tariffs reflect cost of supply and reduce cross-subsidies; and electricity supply should be metered for farmers rather than charged on a per-horsepower basis.[26] As in Maharashtra, most farmers in Andhra Pradesh had opted to pay a supply rate per horsepower rather than for a specific quantum of metered electricity. Agricultural tariffs were thus a central axis of the reform program, particularly critical if the state was to successfully attract private investment to the sector.

Just a few years later, when the reform program began in earnest, however, the Bank's emphasis on the importance of raising agricultural tariffs to the all-around success of the restructuring and privatization program was revised considerably. Similarly, the Andhra Pradesh government under Naidu, who in several public statements had made clear that agricultural subsidies were unsustainable, shielded farmers from any tariff increases in the early years of the program. In a press interview in 1997, for example, Naidu stated that "reforms in the power sector were inevitable. The subsidy on agriculture at the cost of power to industries would one day badly hit the farm sector because there would be little revenue from industries to subsidise farmers . . . [Tariffs] will go up any way, reforms or no reforms. Prices of coal and other inputs have already gone up."[27] Despite these statements at the start of Naidu's reformist agenda, electricity rates for farmers were not increased until after the state assembly election in 1999, and even then were increased under the auspices of the statutorily independent electricity regulatory commission in the hopes that the political class would be shielded from the inevitable popular fallout.

The Bank's lending program for the Andhra Pradesh power sector began with a loan for US$210 million in January 1999. As part of an "adaptable loan," this sum was intended to be the first in a series of five loans that would total US$750 million to fund the much-larger reforms program; subsequent funds were to be released contingent on the state's performance during the preceding period. During the first loan period, the state government was expected to legislate electricity reform; constitute a regulatory commission; and split up the existing utility into separate transmission, generation, and distribution companies that would function as independent corporate entities. The new distribution companies were expected to have gone through at least one round of tariff revisions with the new regulatory body. The 1999 Bank document states that while the agricultural tariff was well below a minimum acceptable threshold (below even the Rs. 0.50/kWh minimum that the central government had arrived at in consultation with states' chief ministers), "it is unlikely that in the short-term, under the current political situation the agricultural tariffs can be increased significantly . . . Once the Regulatory Commission comes into existence, the tariff setting would be depoliticized." The Bank and Naidu were thus explicitly willing to delay raising agricultural tariffs.

There were two main reasons that the Bank and Naidu were willing to relax the requirements for tariff increases for agricultural consumers. Unlike in Odisha, where farmers very quickly faced steep tariff increases, in Andhra, the farmers are a powerful, numerous, and organized political force. Both to prevent electoral losses in the next assembly elections and to deflect popular protest, Naidu understood that delaying tariff increases for farmers would be critical to his longer-term electoral prospects. Both the Bank and the central government were willing to make concessions to Naidu's government because of the critical role he played in the precariously balanced coalition government at the center, where he leveraged his strength not by directly entering the coalition but by providing outside support that he could strategically threaten to withdraw. Other analysts of India's coalition governments as well as the World Bank's state-level lending programs emphasize that Andhra Pradesh received special treatment from both the central government and the World Bank because of the important role that Naidu's TDP played in central coalition politics.[28]

To lay the groundwork for the subsequent restructuring and privatization program, Naidu's electricity sector reforms preceded the World Bank loan by several years. Naidu introduced a bill in the state assembly to restructure APSEB in early 1997. Immediately, the unions responded by declaring an indefinite strike that had no negative impact on power supply in the state. Also immediately, the main opposition parties in the state—including the Congress, the Communist Party of India (CPI), and the CPI (Marxist), but excluding the Bharatiya Janata Party (BJP)—staged a sit-in in the assembly to protest the reform program. Naidu's response in the state assembly was to read from the national Congress Party's election manifesto from the most recent parliamentary elections, in which the party pledged to reform the power sector through the same program of restructuring and privatization that the state unit of the political party was now opposing in Andhra Pradesh.[29] The federal politics of economic reform have meant that many political parties implement economic liberalization in states in which they form the government, yet oppose vehemently the same reforms in states in which they are the opposition. For example, just to the north of Andhra Pradesh, the Congress-ruled Odisha was carrying forward the utility privatization that the previous Janata Dal government under Biju Patnaik had initiated. The same phenomenon characterizes diverging policy stances between the national and provincial units of political parties. The reason that the BJP did not participate in the opposition protest was that by this time, the TDP had already emerged as an important constituent of the federation of parties brought together by the BJP in the national coalition government. In a press conference that followed opposition from political parties and farmers' unions, Naidu pledged that notwithstanding the World Bank's lending agreements, the electricity rates for farmers would not be increased.[30]

The vocal opposition to the reform legislation continued into the next day's assembly meetings, and the members of the Congress, CPI, CPI (Marxist), and several smaller parties were suspended for stalling the assembly's proceedings.[31] In the absence of the opposition, the remaining legislators, mostly Naidu's party members in the TDP, passed the reform legislation without disturbance. The TDP's nearly two-thirds majority in the assembly would have ensured the legislation's passage even had the opposition been present. The legislation accomplished a number of restructuring tasks and paved the way for broad private-sector participation in what in Andhra Pradesh previously had been an entirely government-owned sector. In addition to establishing a regulatory commission, the legislation split up the vertically integrated utility into separate transmission, distribution, and generation companies. The most contentious part of the legislation, and the one that garnered much of the opposition prior to its passage, was the room it left for private companies to enter into all areas of the electricity business.[32] Even more than the sale of state assets to private companies, however, the opposition focused their attention on the World Bank's role in the program, arguing that Naidu had allowed the state's policy to be directed from the Bank's global headquarters.[33] Given the World Bank's involvement with many areas of the state's policy agenda, this antiglobalization, anti-World Bank position became the central thrust of the opposition's political strategy from 1998 to 2004, when the TDP's ten-year reign came to an end.

By the end of 1998, after the government disaggregated the public utility into separate generation, distribution, and transmission companies, and in preparation for distribution privatization, the state government began the necessary tariff increases. Despite Naidu's earlier pledges that farm tariffs were unsustainable and must be increased, the government's tariff increases completely bypassed agricultural consumers. Instead, the rate increases affected approximately 400,000 domestic consumers who used more than a certain quantum of power (protecting the poorest consumers from the upward revision), 300,000 commercial consumers, and large industrial consumers.[34] Both of the latter two categories already paid higher tariffs relative to other states in India, and much higher tariffs than non-Indian commercial and industrial consumers. And despite the government's earlier statements that agricultural subsidies were unsustainable, the bulk of the state's electricity customers, which includes 1.8 million farmers and 6.9 million residences, were completely protected from the changes. In response to the tariff increase, the opposing political parties claimed that the electricity reforms were already proving to be "anti-people."[35] In fact, Naidu and the World Bank had clearly allowed political pragmatism to govern the reforms up to that point. The tariff increases affected only middle-class consumers, industrialists, and small business people, who protested the apparent reversal in Naidu's stance from an earlier unwillingness to increase the burden on industry.[36]

Instead of raising power rates for farmers, the government in 1998 launched a campaign to depress agricultural consumption of electricity. During the sowing season that came just two months before the tariff revision, the government started an agricultural extension program to teach farmers about the benefits of planting crops that are less water-intensive than rice.[37]

Two things informed the government's allocation of tariff increases: the imminent state assembly elections, and the rural protests that had accompanied the last attempt at raising farmers' electricity tariffs. In 1995, shortly after seizing the chief ministership from his father-in-law and founder of the TDP, Naidu raised electricity rates for farmers, a move that elicited large-scale protests and opposition from farmers' groups. The government backed away, and instead paid a subsidy directly to the utility so that the low electricity tariffs for farmers could be maintained. When restructuring and privatization looked imminent, farmers' groups were quick to express their opposition, rightly fearing that the road from restructuring to privatization would require that their subsidized access to electricity be severely curtailed if not eliminated.[38] As farmers understood, no private company would be willing to purchase a distribution system in which the majority of the consumers paid a tariff that was a fraction of production costs.

The upcoming state assembly elections also dissuaded Naidu's government from raising electricity rates for farmers. For Naidu's part, his desire to appear "reformist" to international donors and sell himself within his state as development-minded required that the government impose some tariff increases.[39] However, his need to maintain the populist agenda with which the TDP had become almost synonymous required that the tariff increases spare the largest part of the population, and most important, the rural vote banks. This was possible since the allocation of the tariff increases was subject to the government's discretion, as even the World Bank had agreed that it was not politically feasible to raise rates for farmers.

Electricity politics were likewise a central concern of the main opposition party in the state, the Congress. In the period before the assembly elections, the main emphasis of the Congress Party's agenda was a promise to reinstate free power for farmers. However, because the rate at which farmers were paying was already so low, the promise did not have the intended effect. James Manor has pointed out that another aspect of the regionalization of Indian politics—that is, a shift in the locus of power and importance to the provincial levels—is also reflected in the rhetoric and policies of the two national parties, the Congress and the BJP.[40] He points out that the state units of the Congress in particular come to take on the same political identity as their main regional opponents. In the case of Andhra Pradesh, this has meant that the Congress has countered the populism of the TDP with its own variant of generous consumption subsidies, as their 1999 election agenda demonstrates.

Although there were many other parties in the fray, including the BJP and the various Communist parties, the 1999 election became a two-way contest between Naidu, who projected himself as "development-oriented" and YSR Reddy, the Congress politician, whom Naidu portrayed as regressive. Naidu's TDP captured 180 seats (out of 294) and 43.9 percent of the vote, as against the Congress's 91 seats and 40.6 percent of the vote. As with the electricity reforms, however, Naidu changed policies just enough to comfortably fit the label of a "development-minded politician," but not enough to actually disrupt the system of consumption and production subsidies that benefited the majority of the state's citizenry. There was a carefully cultivated split between Naidu's rhetoric of liberalization and his continuation of populist business as usual.[41] The international business press proclaimed that Naidu's victory signaled that a new style of business-oriented politics was on the ascendant, a statement that exaggerates the TDP's victory (which was only slight) and ignores the actual state of economic policies under Naidu.[42] Many Indian analysts of the elections attribute the TDP's victory to a combination of Naidu's populist development policies and his party's last-minute decision to ally with the BJP.[43]

Just months after the election's end, Naidu held a meeting with his fellow TDP legislators about the power sector. During the meeting, he suggested that tariffs would increase, that it would be their duty as politicians to explain the reason for the increases to their constituents, and that there would be 15 percent tariff increases for each of the next three years.[44] When the increase finally came six months later, it was much larger than any in the political class were prepared for. By this time, an important new actor had emerged, the Andhra Pradesh Electricity Regulatory Commission (APERC). The commission, which came into being in early 1999, took over the tariff-setting responsibilities from the utility and the government in late 1999, although the tariff revision of 2000 was the first under its sole direction. APERC raised tariffs for all classes of consumers, but the rates for domestic consumers increased by the largest amount. Second only to farmers, domestic consumers are the most heavily subsidized class of electricity consumers in most Indian states, including Andhra Pradesh. Unlike the tariff increase imposed in the previous year, which targeted middle-class consumers, this increase affected even the poorest consumers. The average increase was 50 percent, but this climbed to 81 percent for those with the lowest monthly consumption. For farmers, the tariff was increased to 25 paise per unit, which still represented just a fraction of the cost of supply.[45] In providing a rationale for its decision, APERC claimed that with this new set of rate increases, the cross-subsidy burden (on account of farm subsidies) that had been borne exclusively by industrial and commercial consumers would now be shared by domestic consumers as well. In public, Naidu claimed to be as surprised by the tariff increase as everyone else, but asserted that the government was statutorily prevented from interfering in the decision making of the independent regulator.[46] The only way the government could mitigate some of the effects of the increase was to provide a subsidy to the utility and thereby shield domestic consumers from approximately 50 percent of the tariff increase.[47]

APERC's tariff decision, which was implemented by the utility two weeks after it was announced, fomented a long-standing protest led by opposing political parties. Farmers' organizations and civil society groups opposed to the World Bank's involvement in the state joined the protest, but the primary organizers were political actors. Among the tactics employed by the protestors were staging daylong fasts, surrounding and locking up utility staff in regional headquarters and offices, holding rallies and marches, and destroying local utility offices. The political opposition from the Congress and CPI (M) that spearheaded the protests also urged customers to stop paying their bills. Eventually, the sweeping political opposition influenced even Naidu's fellow legislators and party workers in the TDP to oppose the tariff increases. In several general body meetings of the TDP in different districts of the state, power politics dominated the discussion, and the deputy speaker of the assembly, a TDP legislator named A. Chandrasekhara Rao, wrote a public letter urging the chief minister to intervene. In what became an embarrassment to the chief minister, the language of the letter exhorted Naidu to allow the common man to at least hold on to his "langot" or underwear.[48]

The opposition protests continued for ten weeks, until the assembly reconvened in August and the debate was once again focused on the electricity sector. In the assembly, Naidu argued that the government was unable to force APERC to revise its decision, and that the only thing the government could directly do was provide a subsidy to offset the costs to consumers. The government was already doing this for agricultural consumers. In total in 2000–2001, the state government was estimated to pay a subsidy of Rs. 1,626 crores (roughly US$300 million) to the utility so that it could maintain farm subsides. To fully offset the tariff increases for middle-class consumers, the state government would have to provide an additional subsidy of nearly Rs. 800 crores (another US$150 million).[49] The debate in the assembly lasted for almost a week, led to the suspensions of many opposition legislators, displaced assembly debate on other subjects like the state budget, and even led to deaths as police responded with force to a demonstration outside the assembly building.[50] The protest was finally quelled when the chief minister increased the government's subsidy to the utility, which in turn brought down the tariff hike from 20 percent to 14.5 percent, and presented the agreements that the government had signed with the World Bank for the restructuring program to the assembly.[51]

In the time leading to the next assembly elections, however, even the modest tariff increases for agriculture that Naidu's government had managed to implement were rolled back in an attempt to appease the vast farm sector. Naidu promised to deliver nine hours of uninterrupted power supply to all rural areas; the Congress, led again by YSR Reddy, trumped this by again promising free power for farmers.[52] The politics of power played an important part in the elections, although many other factors also led to Naidu's defeat. In a sign of how the Congress Party interpreted its electoral victory, immediately after the election results were declared, Reddy and his new government implemented farmer-friendly policies, including free power and a waiver for unpaid bills. One media outlet reported that "minutes after being sworn in Chief Minister of Andhra Pradesh, Dr. Y. S. Rajasekhara Reddy signed a file in full public glare to give effect to the Congress-I's pre-poll promise of free power to farmers. A little over 23 lakh [2,300,000] farm connections will benefit. This means an annual outgo of Rs. 436 crore [US$86 million] and the one-time waiver will leave the exechequer poorer by Rs. 1,192 crore [US$238 million]."

With this, the first phase of Andhra Pradesh's experiment with electricity privatization ended. The World Bank, in its completion report for what had initially been planned as just the first tranche of a much larger lending project, noted that "subsidized power supply to agriculture is a deeper public policy issue and not a mere sectoral issue," and later, that "unless the distribution business becomes viable and commercial risk of power supply to agriculture is minimized the successful privatization of the [distribution companies] in the present structure would be difficult."[53]

Conclusion

In the end, the Indian politician who was the most vocal advocate of economic liberalization was unable to privatize the electricity distribution business, thanks largely to a state political economy in which significant developmental resources had been extended in earlier periods to provide electricity-based irrigation. More than other cases, Andhra Pradesh under Naidu reveals the extent to which a political will to change is insufficient in the face of conditions that favor the status quo.

In Andhra Pradesh, Chief Minister Naidu, the most ardent advocate of economic liberalization, was unable to counter the built-in resistance to privatization posed by extensive rural subsidies. In Odisha, privatization was achieved with none of the political posturing and maneuverings evident in Andhra Pradesh during Naidu's attempts to privatize his state's distribution infrastructure. The important difference is the lack of subsidies for farmers in the cases of successful privatization and their robust presence in Andhra Pradesh and Maharashtra. The cases of Andhra Pradesh and Maharashtra—in which rural subsidies are an important part of state-level political economies—are more characteristic of the other large states in India, especially those that devoted resources to rural development during the decades of the 1970s and 1980s. The more successful rural developmental states, those that implemented Green Revolution technologies to increase agricultural productivity, were paradoxically locked into this equation of providing subsidies even when they were potentially no longer needed by rural beneficiaries. A state like Odisha, by contrast, skipped over that phase of political economic development that characterized the rest of the country during the 1970s and 1980s. In the absence of the interests that benefited from such subsidy policies, Odisha was therefore able to proceed more easily with the new agenda of economic liberalization and privatization during the 1990s.

6

Conclusion

"Electricity for All"

Chapters 2–5 of this book collectively have two main aims. I begin with the observation that in the creation of modern India, from the end of the colonial period through the first sixty years of independent political life, executive authority in many fields of social and infrastructural development has rested in the hands of state governments rather than with New Delhi. From New Delhi emerged grand themes of national unity—but from the states came the emerging machinery of contemporary India. Tracing the way that India's electricity sector has grown during this period helps uncover the ways that state governments coordinated development in their territories. In Chapter 2, by looking at events, personalities, and agendas in New Delhi from the late colonial period until the early 2000s, I bridge the colonial/postcolonial divide to locate the colonial origins of postcolonial India's institutions. Despite outspoken suggestions that electricity should be organized at an all-India or national level, electricity utilities were made coterminous with state boundaries; the electricity grid took the shape of the federal state. The architecture of electricity that came out of negotiations in New Delhi thereby ensured that patterns of electrification would vary from state to state, and the dream of national unity found no reflection in that architecture. As Chapters 3–5 show, these various state-level patterns of electrification reveal local political economic logics; the expanding grid gave material form to prevailing distributions of political power that often upheld, reinscribed, or even exacerbated rather than subverted rural-urban and intrarural inequalities.

A related goal of this book is to link these regional histories of electrification with events during India's period of economic liberalization. I propose that regional variation in India's power sector reforms from 1991 to 2003 is a function of historical differences in patterns of electrification across the states. To make this argument, I braid histories of electrification across the 1991 divide that has come to separate many political economies of India into the pre- and postmarket reform eras. The manner in which India's state-level political regimes went about building the public electricity sector in their territories from the 1960s to the 1980s constrained and shaped whether and how the successor regimes of the 1990s implemented market reforms. While countries in other parts of Asia and much of Latin America, Europe, and parts of Africa were privatizing public utilities from the 1980s onward, most of India's states either entirely resisted or searched for alternative means of reforming their power sectors that could better accommodate the existing matrix of electricity allocation and pricing.

I conclude with two final aims. The first is to extend the project geographically to consider whether, how well, and with what modifications the arguments of the book help to make sense of political and infrastructural processes in other states in India. In this effort, I briefly consider three additional states—Punjab, Tamil Nadu, and West Bengal. The first two have some of India's highest rates of household electrification, while the last has among the lowest rates of electricity access. According to the 2011 census, in Punjab and Tamil Nadu, the percentage of households that rely on electricity as a primary source of lighting is 96.6 percent and 93.4 percent, respectively; for households in West Bengal, the figure drops to 54.5 percent.[1] The electrical histories of Punjab and West Bengal largely conform to the book's argument: in the postindependence period, rural electrification was most quickly achieved where rural political mobilization redirected development resources to the countryside, and was less likely in its absence. Tamil Nadu demonstrates a different pattern, in which the beginning of rural electrification precedes independence by several decades; consequently, rural electricity access and the use of electricity in agrarian production was much higher than anywhere else in India during the first few decades after independence.

The political dynamics undergirding Gujarat's expanding rural electrification program from the 1960s to the 1980s were very similar to Maharashtra's, and for that reason I do not consider the state in more detail in this conclusion. Conditions in the two states during the 1990s were also similar; rural beneficiaries of electrification organized in Gujarat's most prominent farmer's union, the Bharatiya Kisan Sangh, successfully protected levels of electricity subsidies that were among the highest in the country. But in the aftermath of the 2003 electricity reform legislation, the government in Gujarat headed by Narendra Modi was able to partially break the hold of the peasant union and push through increases in power tariffs for farmers.[2] Gujarat is important, then, in showing that the political dynamics that I highlight in my book are themselves conditional. Modi's Gujarat, as singular as it is among India's states, reveals how a state government that has not just a leader with charisma and support from international agencies (as Naidu did in Andhra Pradesh) but also disciplined party and bureaucratic structures can successfully dismantle a prevailing order.[3]

The second goal of the conclusion is to extend the argument temporally. The empirical scope of this book reaches from the late colonial period until just after the central government passed the Electricity Act of 2003, a piece of legislation that I argue marks the end of a particular era of infrastructural state building in India. In the second half of the conclusion, I very briefly consider the politics of electricity in the decade following the 2003 legislation by looking at a recent moment of crisis. In doing so, I carry forward the discussion from Chapter 2 about center-state dynamics in the debates around India's electricity institutions. The July 2012 blackout that affected much of northern, eastern, and northeastern India suggests that, the new legislation notwithstanding, some of the forces that shaped India's electricity development from independence onward still continue to operate.

Punjab: The Politics of Farm Power

The electrification story in Punjab mirrors in many ways that of Maharashtra, where a strong and organized rural coalition exerted influence over the development ideology and trajectory of the state. The ascendance of a rural political coalition was possibly even more complete in Punjab than in Maharashtra, since in the latter, the rural lobby had to coexist with the country's largest industrial and financial sectors, as well as with one of India's largest private utilities. Additionally, some sociopolitical features of Punjab were entirely unique in the immediate postindependence decades and left their mark as well on the development agenda and trajectory of the state. Most obvious were the events and consequences of the partition of the subcontinent, the violence of which was concentrated in the districts that became part of Indian Punjab and Pakistani Punjab. In addition to immense social dislocation, there were also many economic and infrastructural repercussions of partition, chief of which was that many of the colonial-era investments in canal and well irrigation fell on the Pakistani side of the border.[4] The vast evacuation of lands by communities on both sides of the border meant that on the Indian side, at least, the necessary reordering of land tenure and distribution left a much larger base of medium-sized farmers than in other parts of India.[5] To these structural features of the economy of Punjab after independence, we must add the vital political fact that each postindependence government in Punjab prioritized the development of the agricultural sector almost to the exclusion of all else. This was amply demonstrated in state budgets that included large planned investments on rural infrastructure, early subsidies for agrarian inputs, a steady supply of rural credit, guaranteed minimum prices for farm produce, and extension services to facilitate the adoption of new agrarian technologies.[6]

We get a further sense of how the state in Punjab emphasized agrarian development over other sectors of the economy by looking at planned spending. In the *First Five-Year Plan* period, irrigation and power consumed 58.6 percent of total spending in Punjab, and from then through the *Sixth Five-Year Plan* period, spending on these two never dropped below 50 percent of the total plan. Within spending on power, the government demonstrated an enduring bias in favor of expanding rural electrification and access for agricultural consumption. Unlike the more industrialized parts of British India, the region of Punjab had only modest electricity supplies prior to independence. Until 1960, the per capita consumption of electricity in Punjab was lower than the national average. By 1965–66, the availability of new sources of electricity and a focused effort on rural electrification raised

TABLE 6.1

Electricity consumption in Punjab by category of consumer, 1951 to 1975–1976

	Per capita consumption (kWh)		*Share of electricity consumed by agriculture (%)*		*Share of electricity consumed by industry (%)*	
Year	*Punjab*	*India*	*Punjab*	*India*	*Punjab*	*India*
1951	5.89	17.80	6.90	8.62	54.73	63.82
1956	11.07	26.40	13.73	7.95	54.11	66.89
1960–1961	28.12	38.20	14.47	9.10	61.34	69.50
1965–1966	98.35	61.40	19.24	9.42	63.54	70.60
1970–1971	158.52	89.80	38.00	12.55	35.96	67.65
1975–1976	230.51	110.0	43.16	16.68	35.35	62.36

Source: Government of Punjab, *Statistical Abstract of Punjab*, various years. Cited in G. K. Chadha, *The State and Rural Economic Transformation: The Case of Punjab, 1950–1985* (New Delhi: Sage, 1986), 77.

Punjab's per capita consumption above the national average. More than the quantum of power consumed in Punjab, though, it is the apportionment of power that marks Punjab's electricity sector as unique in India (see Table 6.1).

Thanks to a robust rural electrification program, agriculture accounted for greater electricity consumption than industry, from as early as 1970. Because the state utility had switched from precisely counting to estimating farm consumption in 1968—itself a sign that the state wanted to encourage electricity use on farms—these data might overestimate the actual quantity of agricultural consumption. Farmers in Punjab became increasingly reliant on electrified tube wells; from the 1950s to the 1970s, electric tube wells replaced canals as the most important source of irrigation in the state.[7] The success of Punjab's rural electrification strategy is ironically borne out by the critical depletion of groundwater resources in the state.[8]

As in Maharashtra, in Punjab too, larger farmers benefited more than smaller farmers from these new technologies. This was apparent not only in rates of access to electricity but also in pricing policies. In 1968, almost a decade earlier than in Maharashtra, the government of Punjab implemented a flat-rate method for determining rural tariffs.[9] Instead of relying on a meter reading of electricity consumption, the utility imposed a fixed charge determined by the size of the motor installed on the electric equipment. The decision to mandate flat-rate tariffs for all irrigation pumps privileged larger users much more than smaller farmers, and was even considered punitive by small farmers in some cases.[10] The bias toward large consumers of electricity grew more marked over time. Not just the method of calculating tariffs but also the quality of the supply came to favor larger farmers over small farmers. A 2006 survey of electricity subsidies confirms that larger landholders in Punjab are more likely to own an irrigation pump set, to use electricity rather than diesel, and to express satisfaction with the quality of their electric supply.[11]

One of the main arguments of this book is that underlying the particular trajectories and emphases of development in Indian states are important features of political sociology. The politics of electrification in Punjab bears out this position. The immediate decades after independence witnessed a shift in Punjab from urban-based, upper-caste, and often Hindu elites to agrarian, mostly Jat Sikh farming communities.[12] These communities successfully pushed for measures to increase agrarian productivity, of which a robust rural electrification program was an important and costly component. In turn, the growing availability of electricity in the countryside was one of the reasons that Green Revolution techniques could be so widely adopted.

The economy of Punjab, like that of many other parts of India, has undergone decided structural shifts during the past fifty years. The most significant recent change is the rise of a new service economy, and concomitantly, the declining importance of agriculture. In 1970–71, agriculture contributed 57 percent of the state's gross domestic product; as of 2008–2009, the figure was 26.6 percent.[13] Yet even as Punjab's economy has changed dramatically, exemplified by its burgeoning service sector, subsidies for agriculture remain a dominant feature of the state's political economy. Given the importance of agriculturalists within the state legislature and the importance of subsidies to the largest and most influential farmers in the state, it is perhaps not surprising that successive governments in Punjab have delayed any major reforms to the state's electricity sector that might threaten the links between infrastructural subsidies, political clout, and agrarian production.[14]

Tamil Nadu: "Thinking Electrically" in Colonial Madras[15]

Like Punjab, Tamil Nadu also has had a robust rural electrification program. By 1963, when the Planning Commission surveyed states' rural electrification programs, Tamil Nadu's village electrification rate of 41.6 percent far surpassed all other states in India.[16] Tamil Nadu's more expansive village electrification program was also coupled with a concerted effort to increase agricultural consumption of electricity, which at 22 percent in 1962–63 was much higher than anywhere else. In 1965, not only were tariffs for agricultural consumers in Tamil Nadu lower than the national average, but the state's Electricity Department heavily promoted the use of electricity through short films that depicted how electricity could reduce the costs of irrigation, lessen manual labor in rural households, and increase production in agro-industries.[17] Tamil Nadu was also among the earliest states to establish a "harijan welfare fund" to bring electricity lines to the outskirts of rural towns and villages where formerly untouchable communities were still segregated.[18] In electrified villages in most other states, the grid stopped short of the poorest and most marginalized areas. By 1976, though, the Tamil Nadu State Electricity Board announced that 98 percent of the state's villages and 98 percent of the state's "harijan colonies" were connected to the grid.[19] Likewise, Tamil Nadu was also one of the only states, along with Karnataka and Punjab, to offer financial concessions to promote electrification in "backward" areas of the state with the explicit aim of achieving regional balance.[20] Finally, in Tamil Nadu, much earlier than in other states, the public utility had evolved a system of cross-subsidies that used profits from industrial consumers to offset the losses for rural areas where both the overall demand for electricity and the tariff charged to agricultural consumers were low.[21]

Some scholars of politics in Tamil Nadu have attributed the rapid spread of rural electrification to the success of non-Brahmin politics beginning in 1967. One illuminating study of Tamil Nadu's ethnic politics and populism notes that after the non-Brahmin Dravida Munnetra Kazhagam (DMK) came to power, 21,000 new tube wells were electrified in a two-year span (1967–1969) as part of a larger stable of policies that benefited the middle peasantry.[22] But the records of the Tamil Nadu State Electricity Board tell a different story. Far from beginning under the DMK, the pace of rural electrification had been just as rapid, and even more so under earlier governments. For example, in 1961–62 between 1,000 and 1,500 new irrigation pump sets were connected to the expanding grid every month.[23] The history of planning in the state likewise reveals a firm commitment to the power sector in general and to rural electrification in particular even before the rise of the DMK. For example, during the *First Five-Year Plan*, the power sector alone received 38 percent of total spending, which increased to 41 percent in the second plan period.[24] Unlike in other states, beginning from these early planning periods, there was a twin emphasis on building new generating capacity and on rural electrification.

Unlike in the majority of the country, Tamil Nadu's broad-based electrification program preceded independence by several decades. The explanations that I have offered for rural electrification programs in democratic India—mobilized rural constituencies, state legislatures that championed rural interests, and nodal actors like agricultural cooperatives—cannot make sense of electricity's initial extension to rural Tamil Nadu. For this, we must turn to the colonial period, where the program for rural electrification began.

In 1944, Madras Province had one of the largest electricity systems in British India, measured by generating capacity and total sales. Of the total electricity sold to consumers in the province, 5.7 percent was put to agricultural use. While this may seem very low, it was substantially higher than in the two largest electricity systems of Bombay and Bengal; in Bombay, the agricultural sector consumed 0.02 percent of total electricity, and in Bengal, there was no rural consumption. To explain why there were far greater investments in rural electrification in colonial Madras than other parts of British India, I give a cursory and provisional answer focused on the interplay of provincial politics and public administration.

In the early decades of the twentieth century, authorities in colonial Madras—as elsewhere in the British Empire—were faced with new pressures to increase revenue.[25] With one of the highest land assessments in British India, augmenting revenue through the simple mechanism of raising land taxes was not considered feasible.[26] Instead, the provincial administration sought policies that would expand the sources of revenue rather than just the rate of taxation. Among these were policies to increase the area under cultivation, expand potentially remunerative state activities, and stimulate productive activity by providing new credit facilities.[27] Public investments in electrification fit neatly alongside these endeavors.

Aside from revenue pressures, the nature of provincial administration might provide another explanation as to why public investments in electricity were spatially dispersed rather than concentrated in cities and among industries, as in other parts of British India. Alongside the earlier-mentioned policy reforms, the apex state in colonial Madras also launched a program of administrative devolution. By 1928, when the Royal Commission on Agriculture surveyed and compared the various colonial dominions, the state in colonial Madras was deemed to be highly decentralized and reliant on local institutions for many local administrative functions.[28] Since many of the individuals at the lower levels of administration were drawn from among the landed elites, we might expect that these strata of colonial bureaucracy would have welcomed state initiatives that promised to increase agrarian productivity, such as rural electrification.

As another piece of the story about why Madras embraced electrification, we should also consider the demonstration effect from princely Mysore. There, a successful and profitable state-owned electricity generation and distribution project had shown that by providing this new input—electricity—a state could attract new enterprises and increase the productivity of existing ones. When asked to justify new public investments in hydroelectricity, C. P. Ramaswami Aiyar, a member of the executive council in Madras and one of the project's strongest supporters, pointed to Mysore. He argued that the administration there easily compensated for its public spending through the additional revenues it generated, and added, if Mysore, "why not we?"[29]

Individual champions of technology, like Aiyar, worked from within the highest echelons of the administration to further the project of electrification, and they were supported by other elite and emerging elite groups in the state. Foreshadowing farmer's movements and parties in other parts of India by several decades, the Justice Party, a new elite non-Brahmin political party in Madras that also claimed to represent *ryots*, or peasant proprietors, had begun to champion the cause of rural electrification as early as the 1930s. Seeing the benefits that had already accrued to electrified regions, party activists pledged to support universal access for farmers.[30] Here we might speculate that early non-Brahmin political organizing in Tamil Nadu helped to lay the foundation for a shift from a model of development focused on urban areas and industry to a more inclusive one that could bring the countryside into the equation more fully.

The modalities of electrification in Madras were also quite different than in other parts of British India. Beginning in 1924, Madras Presidency began providing technical and financial assistance to local bodies that wanted to introduce electricity to their territories, collecting fees from the municipalities for their advice.[31] Unlike the rest of the country, then, where private electric utilities held most of the licenses to distribute electricity in towns and cities, in Madras, local governments increasingly took up this job. Furthermore, starting in the 1920s, the provincial government reserved for itself the tasks of transmission and new generation. The decision was motivated by a desire, first, to ensure that electricity produced in different parts of the state was technically well coordinated, and, second, to increase generating efficiency by increasing plant size. The latter motive took Madras's infrastructural development in a wholly different direction than other parts of the country. As discussed in Chapter 3, in Bombay, the Tata conglomerate began developing the hydroelectric potential of the Western Ghats in the early twentieth century. By contrast, in colonial Madras, officials had come to view hydroelectricity as the fitting and exclusive responsibility of the government.[32] The availability of cheap, state-owned hydroelectric power enabled utilities in Madras to charge their customers more affordable rates than in other parts of India.

Many in Madras advocated not just these piecemeal efforts but also a comprehensive electricity policy. In 1930, the chief engineer for hydroelectricity, Major Howard, laid out a rationale for why Madras Presidency needed such an overarching policy. Comparing developments in places as varied as China, the Belgian Congo, and prospects for a Europe-wide electric grid, Howard asserted: "If all these countries have found that electricity is an important economic factor in the lives of the people, then surely, Madras should find it equally advantageous. People must be educated to *think electrically* [emphasis mine] and realize that electricity is just as important to the community as railways, telegraphs, telephones, irrigation systems, and other social necessities."[33] As is hopefully clear from this limited account of colonial electricity history, "thinking electrically" had advanced much further in Madras than elsewhere.[34]

Despite all of the electrical progress in Madras and the exhortations of individuals like Howard, the province did not fully integrate its grid. Instead, the largely hydroelectricity-based grid of the southern districts remained disconnected from the much more modest thermal grids of the north, located in the regions that would become Andhra Pradesh. The result was a spatial unevenness in development, whereby Tamil-speaking regions in the southern parts of the province had access to low-cost hydroelectricity and the Telugu-speaking regions in the north were served by higher-cost thermal power, and that, too, from an electrification process that began at least ten years later. The disparities in access to electricity, among other development issues, exacerbated tensions between the Tamil and Telugu parts of the state, and contributed to the demand for a separate state of Andhra Pradesh.[35]

The successive governments of postindependence Tamil Nadu, then, had a much more substantial infrastructural base from which to expand rural electrification. Likewise, there was a ready embrace of state-owned electricity born of several decades of successful public investment in the sector. Of the constituent assembly members debating electricity in 1948, the representatives from Madras were most voluble in their criticism of private firms for not extending electricity to the farm sector, and most adamant that nationalization would correct this imbalance.[36] Successive governments in Tamil Nadu, from early Congress ministries all the way through the DMK and the All India Anna Dravida Munnetra Kazhagam (AIADMK) administrations, continued to devote substantial sums to the electricity sector.

The effects of such swift and steady electrification on Tamil Nadu's agrarian and agro-industrial development were enormous. As in Maharashtra, the availability of new more-stable and less-expensive irrigation supplies afforded by tube wells shifted the cropping pattern. Overall, there was a substantial increase in rice and sugarcane, both of which require steady irrigation, and a decrease in "dry crops" like native cereals.[37] Given the early and steady use of electricity in agriculture, it is also not surprising that Tamil Nadu was one of the first regions to experience the ecological consequences of a system that exploited groundwater without sufficiently replenishing it. Likewise, Tamil Nadu, where electricity became integral to rural and urban lives and livelihoods much earlier than it did elsewhere, was an early site of electricity shortages that began in the 1970s.[38] These difficulties were compounded by tariffs for agriculture that had already started out lower than in other parts of India, as we saw previously, but that gradually declined to zero in the heated electoral competition between Tamil Nadu's two alternately reigning political parties, the DMK and the AIADMK, from the late 1960s to the present.[39] During the market reform period of the 1990s, it is not surprising that governments in Tamil Nadu never seriously considered the privatization model that was completed in Odisha and attempted in Andhra Pradesh. Instead, the state government attempted various other fiscal and financial reforms, including occasional attempts to reinstate electricity tariffs for farmers. But, as in Punjab, politicians in Tamil Nadu have also found that subsidies for electricity are appealing to put in place but difficult to dismantle.

West Bengal: Redistribution Without Electrification

Compared to Maharashtra and Tamil Nadu, both irrigation pump set electrification and rural household electrification are substantially lower in West Bengal. In this regard, West Bengal resembles its neighbor, Odisha. There too, as discussed in Chapter 4, electricity access continues to be limited and very little is used for agricultural purposes. As in Odisha, the state in West Bengal has done little to expand access to electricity. The relative neglect of the rural sector and rural electrification has persisted from the colonial period to the present, but perhaps most surprising is that the low rates of electrification persisted even throughout the more than thirty year period when West Bengal was governed by a leftist party, the Communist Party of India (Marxist), or CPI(M). With a demonstrated commitment to rural land redistribution and poverty alleviation, we might also have expected more emphasis on measures to expand rural access to public goods such as electricity.

To help explain the patterns of electrification in West Bengal, in what follows I look briefly at the configuration of both the state and state-society relations. The state in West Bengal has not been compelled to increase rural electrification expenditures and agrarian productivity either by rural political coalitions operating from within the state, as in Maharashtra and Punjab, or from organized farmers' movements demanding resources from outside of the state. Instead, the political elites that have governed West Bengal, whether from the Congress Party or the CPI(M), have remained overwhelmingly urban and with a focus on industrial development, despite the evident commitment of the latter to land redistribution and rural equality.

Along with Maharashtra and Tamil Nadu, the region of contemporary West Bengal was one of the oldest and most economically important parts of British India. Also like those two states, West Bengal had a much more substantial industrial economy at independence than the rest of India, and therefore accounted for a major share of the generating capacity and the total electricity produced and consumed within India. The Calcutta Electric Supply Company (CESC) first generated electricity in what was then called Calcutta in 1899.[40] For the next fifty years, the company's operations expanded to cover the domestic consumers of the city, government buildings, and the large textile factories and other industrial undertakings both within and just outside of the city. CESC has continued under private ownership into the present. With the gradual acquisition of CESC beginning in 1989 by the RP Goenka group of industrial companies, the utility came under Indian ownership. Prior to this, though, CESC's status as a foreign-owned utility was the subject of almost continuous critique from the late colonial period onward.[41]

Unlike in colonial Madras province, very little electricity was consumed in rural areas in colonial Bengal, either for lighting or irrigation. After independence, CESC continued to focus almost exclusively on domestic, commercial, and industrial customers, and the task of rural electrification then fell to the newly created public utility, the West Bengal State Electricity Board (WBSEB), which began operating in 1950. At that point, although there was slightly more emphasis on the rural sector, most of the board's budget went toward expanding the generating capacity to feed the rising demand from both industrial and domestic consumers in and around Kolkata. As in Maharashtra, the fact that the bulk of the state's most lucrative customers were served by a private utility meant that West Bengal could not cross-subsidize rural electrification through higher tariffs for other customers. Even the electricity that was produced by the Damodar Valley Corporation—which was shared by West Bengal, Bihar, and central government utilities—was mostly put toward industrial use. In 1977–78, for example, when 82 percent of the DVC's capital expenditure went towards the power sector, more than 50 percent of the electricity it produced was used by industrial units in and around the river valley.[42]

During the colonial period, then, in the relative neglect of rural areas and the ownership of the electricity sector by private rather than state or municipal utilities, the pattern of electrification in West Bengal was more like that of

Maharashtra than Tamil Nadu. In its postindependence history of electrification, though, West Bengal parted ways with both of those early peers. As discussed in Chapter 3, coterminous with the rising power of rural political coalitions, the state in Maharashtra devoted more substantial sums toward rural electrification and rural energy provision, particularly to the rich farmlands that form the subregion of Western Maharashtra. As a consequence of this new development agenda, the productivity, cropping intensity, and wealth of rural Maharashtra increased and, over time, households got access to this new public good as well. As discussed previously, these processes began much earlier in Tamil Nadu, but had many of the same effects.

In most other states, the newly created electricity boards took over private utilities and emerged after several decades of planning as the largest generators and distributors of electricity in their region. In West Bengal, the sheer size and presence of CESC in populous Kolkata and its surrounding industrialized belt meant that the WBSEB grew far more slowly. Fueling some of this difference is the nature of planning in West Bengal. From the start, Tamil Nadu devoted substantial sums to the power sector. Likewise, there was a strong focus on rural electrification and electrifying tube wells with the explicit aim of increasing farm productivity and guaranteeing irrigation resources. A similar pattern governed Maharashtra, particularly from the *Third Five-Year Plan* onward. In West Bengal, by contrast, power and irrigation together took up a smaller overall share of the planning budget, particularly in the early decades of planning. All Indian states taken together devoted an average of 23.1 percent of total planned expenditure to irrigation and power in the first three five-year plans; the figure for West Bengal is 21.5 percent.[43] The contrast is even greater when we compare West Bengal with the most-aggressive rural electrifiers. For example, during the second and third plan periods, despite having one of the country's highest per capita incomes, West Bengal spent Rs. 9 and 17 per capita on irrigation and power, respectively; during the same two plan periods, Maharashtra spent Rs. 18 and 38 per capita, and Tamil Nadu spent Rs. 28 and 37 per capita.[44]

In 1961, fifteen years before the CPI(M) came to power in West Bengal, one of the rising stars of the party was Jyoti Basu. As leader of the opposition, Basu submitted a memorandum to the then Congress government, demanding that the government nationalize the private utility, CESC. The memo urged that "CESC must be taken over by the Government as early as possible, because a private British monopoly concern can no longer be allowed to control the life-line of the nation and hold the people to ransom."[45] The proximate motivation for the opposition's stance was a severe shortage of electricity in the state, which forced CESC to reduce supplies to industry by 40 percent. But after coming to power fifteen years later, the CPI(M) also failed to take over CESC.

Although the upper cadres of the CPI(M) were urban-based, after the party's victory in the 1977 elections, there was a great emphasis on the countryside. First, there was a concerted thrust to politically develop rural areas by strengthening local government, known as the *panchayat* institutions. And second, the new government emphasized rural social sector development and carried out extensive land reform. The consequences of these policies for West Bengal's rural citizens and agrarian economy were enormous. For example, land redistribution, which came to be known as "Operation Barga" (after the *bagirdars,* or sharecroppers, who were its main beneficiaries), greatly expanded the ranks of marginal farmers.[46] Whereas medium and large farms account for two-thirds of cultivated lands in Punjab, in West Bengal, the pattern is inverse. There, it is marginal and small farmers who account for two-thirds of the cropped area.[47]

Under the CPI(M), rural wealth and property redistribution was vigorous; however, the Communist record with respect to increasing agrarian productivity is more mixed. We can consider rural electrification, with all of its attendant consequences for rural lives and livelihoods, to be a key instrument of increasing productivity. Although the government initially tried to augment rural electrification, the policy was too modest and too short-lived to effect a dramatic transformation. For a six-year period, from 1985 to 1991, the WBSEB issued more new connections for electric pumps than had been given out in earlier decades. Even after this increase, though, the rates of rural electrification and the use of electrified tube wells for irrigation remained significantly lower than in other parts of the country. In 1977–78, less than 1 percent of electricity sold in West Bengal was used for irrigation; this figure rose so slowly that by 1985–86, it had risen to only 1.65 percent.[48]

Another way to consider how little electricity is used by West Bengal's farmers is to look at the Indian government's measurement of groundwater usage in each state. The Indian Ministry of Power gives the number of potential electrified pumps for each state in India and then counts how much of that potential has been achieved. As of 1993, when states like Punjab, Maharashtra, and Tamil Nadu had electrified more than 90 percent of their potential irrigation pumps, in West Bengal, only 18.9 percent of the estimated 500,000 pumps that could be accessing groundwater were electrified.[49]

After 1991, even these state efforts were rolled back. Although most rural households and farmers were still without electricity access, there was a steep falloff in rural electrification programs in the state. Instead, the government installed a policy that made getting access to new rural electricity connections extremely difficult. Furthermore, tariffs on electricity for farmers were raised to much higher levels than anywhere else in the country.[50] Together, these two factors have ensured that farmers in West Bengal are far more dependent on expensive diesel pumps to operate their irrigation pump sets than farmers elsewhere in India. In another surprising turn, in the electricity sector, the leftist government welcomed private capital and showed more willingness to embrace market reform than most other states in India.[51] In what Jenkins has called an example of "reform by stealth," the West Bengal government quietly but effectively privatized parts of the state utility by giving a long-term lease of the Kasba gas turbine to the Goenka group.[52]

If West Bengal did not benefit from an early start to rural electrification in the colonial period, as did Tamil Nadu and other parts of southern India, neither did the state pursue rural electrification aggressively in the decades after independence. What is perhaps most surprising given the postindependence political history of West Bengal is that even under the CPI(M) after 1977, the improvements in rural electrification were only marginal. One possible explanation for this lies in the nature of electricity as a commodity and the preference on the part of the CPI(M) for reformist rather than revolutionary policy stances. As Kohli reminds us, the CPI(M) operated within a liberal-democratic context that provided its own set of restraining influences on the party's agenda. Even had leaders of the party preferred a more radical reordering of the state's political economy, their desire to remain in political office meant that they could not risk alienating the state's propertied classes.[53] As party leaders understood, an already stagnating economy could not absorb the shock of more radical policies.

Land redistribution, which was an important part of CPI(M) efforts to alleviate rural poverty, exemplified such a policy: much was accomplished by way of securing the tenure of small farmers without substantially affecting propertied interests elsewhere. Increasing rural electrification would have had a markedly different effect. Developing new electric capacity is a time-consuming and costly project. In the short term, substantially increasing electricity to rural users would have required taking away a resource that is vital to the state's urban domestic consumers and its large industrial houses; the party could not afford to alienate either of these constituencies. This became even more problematic during the 1980s, when West Bengal, like other parts of the country in India, began to face more frequent gaps between electricity demand and supply. As elsewhere, the persistent gap forced the utility to restrict supplies to different consumers; it also led to unplanned blackouts.[54] Under such circumstances, a government wary of alienating industrial and urban constituencies might also opt against a rural electrification program that, while it would serve the interests of equity and social justice, also would unleash even further demand.

As I have found in other states, rural electrification was most successful in states where a sizable presence of rural actors demanded a share of development resources, either from within the state or from the outside through social movement activism. During the last several decades, middle-caste, peasant cultivators have become a substantial presence in legislative houses across India. But, as in Odisha, they are still marginal in West Bengal's state politics.[55] There, the CPI(M) leadership and the elected representatives have tended to be both urban-based and upper caste. In the absence of both rural political coalitions and a political party that explicitly represents agrarian interests to push for more development attention in West Bengal, rural electrification as a way to increase agrarian productivity has remained limited; consequently, rural household electricity access is also lower than elsewhere in the country.

This consideration of the history and politics of electricity in Punjab, Tamil Nadu, and West Bengal is necessarily brief. Nevertheless, it confirms some of the political dynamics found in the longer case studies of this book, for instance the crucial role that a politicized peasantry played in swinging electrification resources to the countryside. Yet, the addition of these analyses also provides a useful reminder that colonial era patterns of infrastructural development continue to resonate powerfully in independent India. For example, the experience of electrification in Tamil Nadu suggests that just as the states in democratic India pursued distinct development policies, so too did the various British provinces and princely states of the colonial period. Such divergence in the colonial era then left the successor political units of independent India with different bases of infrastructural development from which to carry out the gargantuan task of electrification.

Power in the Present

The empirical scope of this book ended with the passage of the 2003 Electricity Act. Since then, enough has happened in the Indian electricity sector to merit another book-length treatment, if not several. For one, there continue to be persistent state-level disparities in access to electricity in India. Variations in historical patterns of electrification, general socioeconomic development, and as I have sought to demonstrate in this book, the political choices made by state governments from the 1950s onward all help to explain unevenness in electrification and electricity access. In the last few pages of the book, though, rather than focus on politics within the states, I want to return to one of the dominant themes of Chapter 2, about how center-state frictions continue to impact the overall electricity sector.

In late July 2012, the electric grid covering a large swath of India—home to roughly 600 million people—collapsed. The power outage, which occurred over two successive days, ultimately touched citizens in all of the states in the northern, eastern, and northeastern regional zones of the country. Trains came to a halt; cars, trucks, and buses clogged highways and roads as traffic signals went dark; and hundreds of millions of households, shops, manufacturers, and hospitals lost supply from the electric grid. The lucky among those turned to backup power sources, either diesel generating sets or power invertors. Many residents in the affected regions were spared the worst effects of the blackout for an altogether different reason—because they are among the millions of Indian households that still have no electric connection or get their electricity "off-grid" through highly localized initiatives, such as the one portrayed in the film *Swades*. The event garnered front-page attention in international and Indian news outlets alike.[56] Many analysts noted the irony that a country aspiring to superpower status, home to a sophisticated, high-technology sector, still fails to deliver what is considered a basic and essential infrastructural good.

While the crisis was still unfolding, the central government appointed a three-member commission to investigate the causes of the power failure and suggest remedies. The commission prepared its final report just a few weeks later.[57] The report catalogs a series of technical failures that both sparked the initial grid collapse and allowed it to spread to an ever-larger region. The list of proposed solutions to these technical problems includes the following: expand transmission infrastructure both between and within states; implement methods to "ring-fence," or isolate, discreet parts of the grid in case of electrical failures; upgrade the systems to monitor and communicate about power flows; and reduce the time it takes for failed generating stations to resume operating.

In addition to these technical failures, the report also points to a set of ongoing institutional failures behind the blackout that indicate the complexity of India's electric supply system. Since 1947, the Indian power sector has ballooned in terms of the total capacity of the sector and the consumers it serves. The number and configuration of government departments, utilities, companies, regulators, and system operators that populate it has likewise increased. Coordination failures within this tangled matrix of organizations at the state, regional, and national levels contributed to the 2012 blackout. As the various regional grids were progressively integrated from the 1970s onward, a host of new organizations were created to ensure that power flowing into and out of one grid or region remained stable, veering neither too far below nor above the expected frequency. These new organizations are referred to as system operators and include the National Load Dispatch Centre (NLDC), the five regional load dispatch centers, and the plethora of state load dispatch centers. For an integrated national grid to function smoothly, each of the state load dispatchers must consider the overall viability of the system rather than the supply of power within their own state's borders. If either too much or too little power is taken at any one node in an integrated system—or, conversely, if too much or too little power is pumped into the system—the result can be a grid collapse, which can spill over from one region into the next. The job of the regional load dispatch centers is to monitor the power flows in real time and ask the constituent state load dispatch centers when necessary to adjust the amount of power they are taking from or putting into the system. Instead, the requests made by regional load dispatchers are often ignored, or honored only halfheartedly. This is part of what happened across northern, eastern, and northeastern India in 2012. The committee investigating the blackout listened to recordings of phone conversations between the regional load dispatcher and state dispatchers in its territory. The former made "desperate efforts" to convince state dispatchers to reduce the amount of power they were drawing from the system, often to no avail.[58]

The state load dispatch centers, as it turns out, are under tremendous pressure to accomplish state-level agendas, even at the risk of system-wide failures. The perverse incentives can run in both directions: on the one hand, sometimes states overdraw power from the system to satisfy demand for electricity in their territories, and at other times, states "underdraw" in order to reduce their financial liabilities. Both strategies have negative consequences for the stability of the larger system. This problem was recognized in New Delhi among central power ministries and regulators even before the blackout. In 2008, the Ministry of Power commissioned a study on the functioning of India's load dispatch operators.[59] The committee's three main recommendations were: (1) to improve employees' technical qualifications (through higher salaries and continuing education); (2) to ensure that dispatchers could function autonomously (by being separated from the state utilities and receiving separate funding); and (3) to enhance coordination among the more than three dozen system operators. The report on the 2012 blackout echoed these recommendations, and additionally pointed out that the 2003 Electricity Act needed to be amended so that state-level actors that do not comply with regional and national directives would be appropriately punished. As the report notes, "non-compliance of instructions" made by the regional dispatchers has been a long-standing problem. Further, the "maximum penalty that can be imposed . . . is meager in comparison to the damage that such non-compliance can cause the grid."[60] And even the small penalties that regulators are entitled to impose are often ignored and left unpaid.

Again, we see that the central government and its agencies have a vision for the Indian electricity sector as a whole that is countermanded by state-level actors who work to pursue their own ends. In this regard, we can hear echoes of the 1940s. Ambedkar, for example, was worried that water and coal would not be adequately shared if states were only concerned about providing for their own populations, and he and others proposed that the nature of electricity required greater centralized control. The problem Ambedkar identified remains the same, although it now takes on a much more complex organizational form.

Aside from spectacular but episodic failures like the 2012 blackout, the more-telling evidence that the current matrix of institutions and incentives in the electricity sector is not working for everybody comes from data on household access. Just two weeks after the blackout in 2012, Prime Minister Manmohan Singh gave a speech to celebrate India's sixty-third Independence Day, in which he promised that all Indian households would have electricity within five years' time.[61] In the speech, he did not allude to the election promise that his party had made nearly a decade earlier, which was to provide "electricity for all" households by 2009. The 2011 census demonstrates that for all of the gains made in Indian electrification from independence to the present, electricity is still not a basic public good, available to all citizens. Only 67.2 percent of households can rely on electricity as a primary source of lighting; everyone else—close to 400 million Indians—relies on other sources of fuel, like firewood and kerosene.

Certainly, electrification is one of the most capital-intensive and technologically intricate goods to provide. But other countries have fared better than India. Consider China, where the government estimates that only 1 percent of households are without electricity access.[62] If the remaining hundreds of millions of Indian households are to be electrified and if electricity is to flow continuously, central and state governments need to ensure that the proliferation of agencies, regulators, and utilities are able to coordinate more effectively than they have done up to now. In this regard, the 2003 Electricity Act perhaps did not go far enough. Critics of the legislation argued that the main point of the act was to privatize the power sector by encouraging ever-greater flows of private investment. What is perhaps more necessary, however, is to reform the governance structures of the power sector to give more authority to regulatory institutions and utilities that have incentives to consider a more regional and national perspective in making decisions about how electricity is allocated and priced. Indian citizens also need to demand a more just distribution of resources. In contemporary discussions of what ails India's power sector, the blame is often laid at the doorstep of farmers. But as we see in previous chapters, relatively few farmers benefit from subsidies and the wealthiest gain the most. To finish the project of *electrifying India* will require both political mobilizations from below—by those who were left out of the first wave of infrastructure expansion in the countryside—and institutional restructuring from above.

Notes

Chapter 1

1. The film's original title, *Swades: We, the People,* uses a combination of Hindi and English words, exemplary of how languages intermix in India.

2. With few exceptions, much of the writing on the political and social conditions of electrification focuses on North America and Europe. A landmark history of electrification is Hughes, *Networks of Power.* Chapters 7–9 compare the relationships between technology and politics in three Western locales—Berlin, Chicago, and London. On electricity's impact in the United States, see Nye, *Electrifying America,* the title of which inspired my own. On how society and culture impacted the emergence of public gas and electric systems and were affected in their turn in nineteenth-century Europe, see Schivelbusch, *Disenchanted Night.* See also a pioneering political economy of electricity in Brazil that shows how a mixed public and private sector flourished despite capital constraints thanks to technological characteristics of the sector: Tendler, *Electric Power in Brazil.* A very recent addition to this literature persuasively charts the links between forms of political power and energy systems: Mitchell, *Carbon Democracy.*

3. Nye, *Electrifying America,* 140.

4. In 2011, the spelling of Odisha was officially changed from Orissa. I use the new spelling throughout this book, except when the old spelling appears in direct quotes or titles.

5. Przeworski, *Democracy and the Market.*

6. Schamis, *Re-forming the State.*

7. Weyland, *Politics of Market Reform.*

8. The scholar most closely associated with this idea is Waterbury, *Exposed to Innumerable Delusions.*

9. On the importance of ideology in pension privatization, see Madrid, *Retiring the State.* On how ideology influenced the modalities of privatization in Eastern Europe, see Appel, "Ideological Determinants of Liberal Economic Reform," 520–49. For a related

argument that uses the frames of partisanship and party affiliation to explain regulation strategies in postprivatization utility markets in Latin America, see Murillo, *Political Competition, Partisanship, and Policy Making in Latin American Public Utilities.*

10. Weyland, *Politics of Market Reform*, draws on psychological insights about risk aversion and tolerance, and emphasizes the intersection of new political leaders and a catalyzing economic crisis.

11. On the centrality of economic crisis, see Bhagwati, *India in Transition*; and Jalan, *India's Economic Crisis.* On the importance of ideology in the Indian case, see Rahul Mukherji, "India's Aborted Liberalization," 375–92.

12. Jenkins argues that market reforms in India during the 1990s were carried out most often and most successfully by "stealth." Jenkins, *Democratic Politics and Economic Reform.*

13. If we take the early 1990s as a base, then by a host of metrics, the Odisha State Electricity Board (OSEB) was in the middle of the pack. For example, from 1992 to 1995, the OSEB had a ratio of between five and six employees for every million kilowatt-hours (MkWh) of electricity it produced, comparable to the ratios in Himachal Pradesh, Haryana, and Bihar. Over the same period, it cost less to produce a unit of electricity in Odisha than in Maharashtra, Tamil Nadu, and Gujarat, a fact that was reflected in lower-than-average tariffs across all categories of consumers in Odisha. Odisha had a more attractive mix of hydro and thermal power resources than other states, which lowered the costs of generation. By other measures, too (e.g., sales revenue as a ratio of cost, commercial profit/loss, rate of return), the OSEB was no worse than other SEBs, and by some measures, even better. For comparative data on the technical and financial performance of the SEBs, see Planning Commission, "Annual Report on the Working of State Electricity Boards and Electricity Departments."

14. Ludden, "India's Development Regime," 269.

15. Mann, "Autonomous Power of the State," 185–213.

16. Ibid., 113.

17. O'Donnell, "On the State, Democratization, and Some Conceptual Problems," 1360.

18. Constituent Assembly of India, *Debates*, III, No. 2 (December 12, 1947), 1829. Quoted in Franda, *West Bengal and the Federalizing Process in India*, 69.

19. Soifer and vom Hau, Special Issue, *Studies in Comparative International Development.*

20. Ziblatt, "Why Some Cities Provide More Public Goods than Others," 275.

21. Mann suggests that the growth of infrastructural power during the twentieth century is eclipsing differences between subnational regions in the United States. By the end, "Alabama is more similar to California in providing infrastructures than either is to Mexico." Mann, "Infrastructural Power Revisited," 358. In his introduction to the special issue, Hillel Soifer departs somewhat from Mann. He delineates three distinct research strategies that could draw from Mann's concept. While the first two hinge on the concept of the central state, the third is an explicit engagement with the "uneven reach of the state." Soifer, "State Infrastructural Power," 242.

22. Mann, "Infrastructural Power Revisited," 357–58.

23. In paying closer attention to how state power is structured on the ground, I was influenced greatly by a body of scholarship on state-society relations that has grown during the past twenty-five years. See, for example, Migdal, Kohli, and Shue, *State*

Power and Social Forces. On spatial variation in state-society relations and forms of state power in the context of Africa, see Boone, *Political Topographies of the African State.*

24. For example, see Brass, *Politics of India Since Independence*; and Dasgupta, "India's Federal Design."

25. Bednar, *Robust Federation.* Bednar's three conditions for federal systems include (1) geopolitical divisions that cannot be abrogated unilaterally, (2) independent bases of authority, whether through elections or other means, and (3) policy sovereignty in specific domains and direct governance by each level of authority. India fails to satisfy the first condition since the national parliament can redraw state boundaries unilaterally. Bednar concedes that some scholars of federalism would prefer to look at how federal arrangements play out in practice rather than impose a formal definition.

26. Dasgupta, "India's Federal Design," 63n15, lists ninety-nine subjects under the central list, sixty-one on the state list, and fifty-two on the concurrent list. In cases of overlap on the last list, the central legislation is paramount.

27. Rudolph and Rudolph, "Federalism as State Formation in India," 553–72.

28. Ibid., 565.

29. Rudolph and Rudolph, "Iconization of Chandrababu," 1541–52. Saez, too, agrees that market reforms have strengthened the hand of state governments, but suggests that changes in the party system during the 1980s played a critical role as well: Saez, *Federalism Without a Centre.*

30. Austin, *Indian Constitution,* 188–92.

31. Jalal, *Democracy and Authoritarianism in South Asia.*

32. Ibid., 253.

33. Bardhan, *Political Economy of Development in India,* 40.

34. Rudolph and Rudolph, *In Pursuit of Lakshmi.*

35. Kaviraj, "Critique of the Passive Revolution," 2429–44. Like Bardhan, Sudipta Kaviraj departs from a textbook Marxist perspective in that the state in his theory is accorded an autonomous range of motion. For both scholars, the state is one member of a tripartite dominant coalition. But the state in both views is treated as a singularity.

36. Frankel, *India's Political Economy,* 5.

37. Recent scholarship includes Bussell, *Corruption and Reform in India*; Kirk, *India and the World Bank*; Kohli, *Poverty amid Plenty*; and Singh, "We-ness and Welfare." Some of this work compares differences not only within India but across other countries in South Asia that once formed part of the British Empire. See for example, Herring, *Land to the Tiller*; and Teitelbaum, *Mobilizing Restraint.*

38. Sinha, *Regional Roots of Developmental Politics in India,* 33.

39. Kohli, *State and Poverty in India.*

40. Desai, *State Formation and Radical Democracy in India.*

41. Herring and Agarwala, "Introduction: Restoring Agency to Class," 323–57; Chibber, *Locked in Place.*

42. Harriss, "Comparing Political Regimes," 3367–77.

43. Brian Min makes a related argument, although focused specifically on how electoral politics influence public goods provision. He finds an increase in the provision of electricity to constituencies populated by *dalit*, or former "untouchable," communities when the lower-caste government of Mayawati's Bahujan Samaj Party was in office in Uttar Pradesh during the 1990s. Min, "Distributing Power: Electrifying the Poor in India."

44. The urban bias thesis was first formulated by Lipton, *Why Poor People Stay Poor*, and extended by Bates, *Markets and States in Tropical Africa*.

45. A special issue of the *Journal of Development Studies*, titled *Beyond Urban Bias* and edited by Ashutosh Varshney, was devoted to scholarship that challenged the urban bias thesis with fresh evidence and included responses by Lipton and Bates.

46. Varshney, *Democracy, Development, and the Countryside*.

47. For an early critique of Lipton's focus on rural-urban conflict to the exclusion of all other social conflicts, see Byres, "Of Neo-populist Pipe-Dreams," 210–44.

48. Snyder, "Scaling Down," 93–110.

49. Chibber makes a similar argument for looking again at the 1950s in New Delhi as the decade in which the central apparatus and institutions first took shape. As he argues, the choices and battles of that seminal era conditioned New Delhi's efforts at fomenting an industrial transformation in later decades. For a concise statement of the argument see Chibber, *Locked in Place*, 7–12.

50. Guha makes these points in the prologue to his history of independent India. Guha, *India After Gandhi*, xxii–xxiii.

Chapter 2

Portions of an early version of this chapter appeared in *Pacific Affairs* 77, no. 3.

1. Goswami, *Producing India*, 103.

2. Naoroji, *Poverty and Un-British Rule in India*. Naoroji uses the railways throughout his book as an example of how the benefits of public works accrued in the main to the British.

3. Prakash, *Another Reason*, 164–65.

4. Headrick, *Tools of Empire*, 184–85.

5. See Goswami, *Producing India*, chaps. 1 and 3, which include lengthy excerpts from colonial-era letters and bureaucratic circulars that attest to this.

6. For a longer discussion on my use of Mann's term, see Chapter 1.

7. Nehru seemed keenly aware of this function of public works, as scholars of the early years after independence make clear. See, for example, Khilnani, *Idea of India*; and Roy, *Beyond Belief*. Roy in particular focuses on the role of India's new urban spaces and the role of science and technology in authorizing the political agency of the new nation-state.

8. Mitchell, *Rule of Experts*, 21.

9. For an account of the sensory aspects of Indian nationalism across a range of venues, including state-funded documentary films, see Roy, *Beyond Belief*.

10. Gandhi, "Economic Development," 158.

11. For a discussion of how centralized systems like water and electricity grids eclipsed an older model of household autonomy, see Schivelbusch, *Disenchanted Night*, 28.

12. Gandhi, "Mass Production." Cited in Vinaik, *Sarvodaya and Electricity*, 8–9.

13. Although Gandhian philosophies of development continue to be marginalized today, several different critiques of dominant models of development can claim an affinity with Gandhi's views. The most obvious of these might be environmental movements that target mainstream development via industrialization as an ecologically unstable and unsustainable path for India.

14. Parekh, "Nehru and the National Philosophy of India," 35–48; Singh, *Jawaharlal Nehru on Science and Society*.

15. Nehru, *Glimpses of World History*, 848.

16. Ibid.

17. Chadaga, "Light in Captivity," 82–105.

18. Coopersmith, *Electrification of Russia*, esp. chap. 6, "GOELRO: The Creation of a Dream, 1920–1921." Coopersmith attributes the idea of "technological utopia" to Segal, *Technological Utopianism*, 14.

19. Coopersmith, *Electrification of Russia*, 151.

20. Visvesvaraya's trips outside of India, where he visited sewage and water systems, hydroelectric dams, irrigation networks, and institutes of higher and technical education, inspired new ideas of progress. As he writes in his autobiography, "my travels and studies in Europe, Japan and America had filled my mind with the great possibilities of progress and prosperity which lay before our country." Visvesvaraya, *Memoirs of My Working Life*, 76; see also references to his travels on 42–43, 46, and 121–30.

21. For an account of the cultural foundations of Mysore's early developmentalism, see Gowda, "'Advance Mysore!'" 88–95.

22. Visvesvaraya, *Memoirs of My Working Life*, 78–79.

23. Ibid., 46–47.

24. Visvesvaraya, *Planned Economy for India*.

25. Ibid., 86.

26. Hughes, *Networks of Power*, 350–53.

27. For a detailed excavation of the early history of planning in India, see Chattopadhyay, "Idea of Planning in India."

28. The quote is taken from the title of a recent book that looks at Ambedkar's significance in Indian political and social life: Anupuma Rao, *Caste Question*.

29. Ambedkar, "Post-War Development of Electric Power in India," 280.

30. Ibid.

31. Ambedkar, "Post-War Electric Power Development."

32. World War I had played a similar role in motivating European states to take charge of electricity. See Hughes, *Networks of Power*, 296, 324.

33. "Electricity Economic Index: New Feature of Monthly Business Survey," *Indian Information Series*, February 15, 1939 (New Delhi: Government of India, Bureau of Public Information), 78.

34. "Roger Mission's Recommendation: Electrical Commissioner Appointed," *Indian Information*, March 1, 1941 (New Delhi: Government of India, Bureau of Public Information), 134.

35. B. R. Ambedkar, during a speech made to representatives of the central, Bengal, and Bihar governments at a conference to discuss the Damodar Valley; Bengal Secretariat in Calcutta, January 3, 1945. Reprinted as "Dr. B. R. Ambedkar on Damodar Valley Scheme, Calcutta Conference Address," *Indian Information*, February 1, 1945.

36. This description of the village was part of Ambedkar's response to those who, during the Constituent Assembly debates, advocated devolving greater executive, legislative, and judicial authority to the village level. Quoted in Jodhka, "Nation and Village," 3351.

37. "Development of Sone River Valley: A Multi-purpose Project," *Indian Information*, April 1, 1945.

38. Recent books that touch on this are: Klingensmith, *"One Valley and a Thousand"*; D'Souza, *Drowned and Damned.*

39. "The Story of the Tennessee Valley Authority," *Indian Information*, May 15, 1945.

40. For insightful biographic details about Meghnad Saha and his long association with the Damodar Valley Authority (DVA), as well as the relationship of the latter to the US government's Tennessee Valley Authority (TVA), see Klingensmith, *"One Valley and a Thousand,"* 109–54.

41. Ibid., 131.

42. National Planning Committee, *Report of the Sub-committee on Power and Fuel.*

43. Quoted in D'Souza, *Drowned and Damned*, 188.

44. Electricity (Supply) Act, 1948, Act No. 54 of 1948, preamble.

45. Constituent Assembly Debates (CAD), vol. VI, part II, August 9–31, 1948, 54.

46. Ibid., 41.

47. Ibid., 332.

48. Taylor, *India: Economic Issues in the Power Sector*, 5, and data from annex B2.1. The report is a summary of findings from a World Bank mission to India in 1978–79.

49. CAD, 50.

50. The select committee on the Electricity (Supply) Act had extensive discussion with private utilities, the constituency who would be most affected by the creation of State Electricity Boards (SEBs). The private representatives included Manu Subedar, who spoke on behalf of the Federation of Indian Chambers of Commerce; representatives of the Federation of Electricity Undertakings of India; and J. D. Choksi and W. S. Nelso, who together represented the Tata Hydro Electric Company, which was then the largest private utility in the country. The private sector as a whole favored more central governmental control, worried that leaving the administration of the sector to the provinces would result in taxation rates and rules of operation that differed greatly from region to region. The representatives from private firms also were concerned that since the state governments would own the SEBs, giving those same governments the authority to regulate the private sector would ensure a conflict of interest.

51. In the 1948 Constituent Assembly Debates about the new electricity legislation, representatives from Punjab (C. R. Singh), Madras (T. T. Krishnamachari and M. A. Ayyangar), and Mysore all argued that the provincial governments should continue to be the primary beneficiaries of taxes on electricity. Madras had just imposed prohibition, and Punjab was planning to do so imminently, so representatives from these states were particularly concerned about making up these revenue shortfalls. CAD, 320, 331, and 338.

52. Ibid., 61.

53. Ibid., 62.

54. During the legislative debates in 1948, N. V. Gadgil, who was then minister for works, mines, and power, commented that the government had begun to draft the legislation in 1942: CAD, 332. In January 1948, during meetings of the select committee with members of the privately held electric utilities, one legislator commented that the earliest drafts for the bill were a product of the government under Viceroy Wavell, who held that post from 1943 to 1947. *Select Committee on the Electricity (Supply) Bill, 1947* (New Delhi: Government of India Press, 1948), 31.

55. Ibid., 27.

56. Austin, *Indian Constitution*, 200–201.

57. Appleby, *Public Administration in India*, 42.

58. Appleby, 40.

59. In 2011–12, nuclear energy accounted for 2.7 percent of total electricity in India. Of the balance, 65.3 percent was produced using thermal sources, mostly coal; 21.5 percent from hydroelectric plants; and 10.4 percent from renewable sources. Ministry of Power, *Annual Report 2011–12* (New Delhi: Government of India, 2012), 49, http://powermin.nic.in/reports/pdf/Annual_Report_2011-12_English.pdf (accessed July 24, 2012).

60. Hanson, *Process of Planning*, 205. During the *Fifth Five-Year Plan*, 1974–79, state governments collectively spent more than 85 percent of the total funds for the power sector. Cited in Pachauri, *Energy and Economic Development in India*, 147.

61. Section 10.5, which was added by Act 101 of 1956, states: "If the Board fails to carry out its functions, or refuses or fails to follow the directions issued by the State Government under this Act, the State Government may remove the Chairman and the members of the Board and appoint a Chairman and members in their places."

62. King, *Economic Development Projects*, 229–31.

63. Venkataraman, *Power Development in India*, 60–61.

64. *Report of the Power Tariff Policy Committee* (Venkataraman Committee Report) (New Delhi: Government of India, 1964), 9.

65. Electricity (Supply) Act, 1948, Section 18(a) states that the boards must arrange for "the supply of electricity . . . and for the transmission and distribution of the same in the most efficient and economical manner with particular reference to those areas which are not for the time being supplied or adequately supplied with electricity."

66. Planning Commission, *Report on the Evaluation of the Rural Electrification Programme* (New Delhi: Programme Evaluation Organisation, 1965), 32.

67. *Report of the All-India Rural Credit Review Committee* (Bombay: Reserve Bank of India, 1969), 819–30.

68. Tamil Nadu is an outlier in terms of its rural electrification program and in its electricity sector as a whole, which I discuss in Chapter 6.

69. Ministry of Irrigation and Power, *Report of the Power Economy Committee* (New Delhi: Government of India, 1971), ix–x. See also Ministry of Irrigation and Power, *Power Economy Committee, Report of Study Group, Vol. I* (New Delhi: Government of India, 1973).

70. Venkataraman, *Power Development in India*, 160–61.

71. Henderson, *India: The Energy Sector*, 163.

72. Ibid., 164.

73. Ibid.

74. An early example illustrates this point: In April 1963, an electricity shortage reduced yarn production in Tamil Nadu, which in turn constrained production in the weaving industry located in Nagpur and Malegaon, Maharashtra. A neat triangular trade provided at least a short-term solution: Tamil Nadu supplied yarn to Maharashtra. The latter sent electricity to neighboring Mysore, which in turn sent corresponding electricity to Tamil Nadu. Bhanage, "Regulation and Economics," 205.

75. Chatterjee, *A Possible India*, 67–73; Corbridge and Harriss, *Reinventing India*, 78–83; Charan Singh, *India's Economic Policy*, 47.

76. Sur and Umali-Deininger, "Public Expenditures."

77. Several books and articles explore this issue. On the emergence of water markets in Gujarat, see Dubash, *Tubewell Capitalism*.

78. According to the documents I have consulted, Punjab was the first state to implement flat-rate tariffs, beginning in 1968. I discuss this further in Chapter 6.

79. Power and Energy Division, "Annual Report on the Working of State Power Utilities and Electricity Departments, 2011–2012" (New Delhi: Planning Commission, 2011). From 1980 to 2011, captive power capacity grew from an insignificant amount to nearly 30,000 megawatts (MW); see fig. 1.7 on 8.

80. Corruption is alluded to broadly in the press, government reports, and conversations with people connected to the industry, but to my knowledge, there has been no systematic study of employee-aided corruption in the electricity sector. On corruption in a related sector—canal irrigation—see Wade, "System of Administrative and Political Corruption," 287–328.

81. For example, N. S. Vasant, a former SEB chairman, stated that in a particular northern state, the entire list of contractors hired for projects in the electricity sector changed when there was a change in the party in power; interview by author, New Delhi, October 25, 2002.

82. Tongia emphasizes the fiscal crisis as among the most important motivations for the central government's power sector reforms: Tongia, "Political Economy of Indian Power Sector Reforms," 136–37.

83. For an analysis of the global energy paradigm shift, see Dubash, *Power Politics*, esp. chaps. 1 and 2.

84. Lok Sabha Debates, 1st sess., 10th ser., vol. 5, no. 48, September 17, 1991, 76–135.

85. Ibid., 81.

86. Ibid., 87.

87. *Economic Survey of India, 1980–81* (New Delhi: Ministry of Finance, 1981), 14–15.

88. The commercial losses of all SEBs combined amounted to just over Rs. 4,000 crore in 1990–91, a sum that doubled by 1995–96. *Economic Survey of India, 1995–96* (New Delhi: Ministry of Finance, 1996), 163.

89. Manibog, Dominguez, and Wagner, *Power for Development*, 2.

90. Since the mid-1990s, some have suggested that managerial rather than structural reforms would solve most problems in the power sector, a view echoed by unionists in the sector and antiprivatization activists. See Ruet, *Against the Current*; and Kannan and Pillai, *Plight of the Power Sector in India*.

91. Widespread labor opposition to electricity reform did not begin in earnest until the late 1990s, when states began to dismantle the SEBs. The National Coordination Committee of Electricity Employees and Engineers was formed only in 2000, when labor protests in Uttar Pradesh followed the state's announcement of plans to unbundle its SEB. B. S. Meel, general secretary, Electricity Employees Federation of India, interview by author, New Delhi, December 5, 2002.

92. Ministry of Power, *Annual Report 1991–92* (New Delhi: Government of India, 1992), 29, http://powermin.nic.in/reports/pdf/ar91-92.pdf (accessed July 25, 2012).

93. *Gazette of India*, no. 237, New Delhi, October 1991.

94. Sarkar, "Letter from India," 66.

95. There is an extensive critical literature about the Dabhol project. See Mehta, *Power Play*; and Purkayastha and Prashad, *Enron Blowout*. The journals *Economic and Political Weekly, Frontline, Outlook, and Business India* have devoted many articles to the issue, as have all of the Indian financial and general newspapers.

96. Mehta, *Power Play*, 27–30. See also "Enron Backtracks, Deal Nears," *Financial Times, FT Energy Newsletters*, September 13, 1993.

97. "The Enron Corporation: Corporate Complicity in Human Rights Violations," Human Rights Watch, 1999; "The 'Enron Project' in Maharashtra—Protests Suppressed in the Name of Development," Amnesty International, 1997.

98. Whereas a monopoly implies a single seller and multiple buyers, a monopsony is a structure in which multiple producers must sell to a single buyer.

99. One particularly harsh criticism of regulation in Delhi that questions the commission's insulation from both government and private companies alike is made by Haldea, "Crisis of Credibility."

100. Phadke and Rajan, "Electricity Reforms in India," 3061–72.

101. Prayas Energy Group, "Good Beginning," 41–45.

102. Ashok Rao, member of National Working Group on Power, interview by author, New Delhi, October 9, 2003.

103. See, for example, Purkayastha, "Electricity Bill 2003."

104. This point was made to me during several interviews, particularly by critics of privatization as the solution to the sector's problems. Prabir Purkayastha, who was then with the Delhi Science Forum, made the point most succinctly; interview by author, New Delhi, October 23, 2002.

105. I was able to attend Parliament as a visitor on the day that the Electricity Act was debated in 2003. The brief debate of just about one hour was a striking contrast to the hours of animated and passionate discussion in the Constituent Assembly in 1947 and 1948.

106. Haldea, "Empower Consumers."

107. "Empowering Contradictions in the Electricity Act," *Financial Express*, July 27, 2003.

108. Ministry of Power, *Annual Report*, various years.

109. Ministry of Power, *Annual Report 2002–03*, 15.

Chapter 3

1. In the 2010–11 crushing season (the latest for which I could find data), just five districts, all in Western Maharashtra—Pune, Sangli, Ahmednagar, Kolhapur, and Solapur—produced 70 percent of the state's sugar output. See data from the Vasantdada Sugar Institute, http://www.vsisugar.com/india/statistics/maharashtra_statistics.htm (accessed August 31, 2012).

2. Sirsikar, *Politics of Modern Maharashtra*. The late Rajendra Vora observed that the Shiv Sena's successes from the 1990s onward represented a transfer of political power from the countryside to the city: Vora, "Shift of Power," 171–73.

3. Rosenthal, *Expansive Elite*, 13.

4. The city's name was changed from Bombay to Mumbai in 1995.

5. Samyutka Maharashtra Parishad, *Reorganization of States in India*, 5–6.

6. For the first several decades, the conferences took place every few years, but became annual events from 1927 onward. For a cultural history of the Samyukta movement, see Palshikar, "Breaking Bombay, Making Maharashtra."

7. Samyukta Maharashtra Parishad, *Reorganization of States in India*, 13.

8. As cited in Phadke "Samyukta Maharashtra Movement," 118–34.

9. Ambedkar, *Thoughts on Linguistic States*. In the preface of what appears to be a self-published booklet, Ambedkar notes that illness had kept him from the parliamentary debate on linguistic states. After seeing the report of the States Reorganisation Commission, he was prompted to offer his analysis on the subject.

10. Unlike most of the leadership of the Samyukta movement, of which he was not a part, Ambedkar advocated splitting Bombay Province into four parts: Maharashtra City State (comprised mainly of metropolitan Bombay) and Western, Eastern, and Central Maharashtra. This likely reflects his view of the further inequities in the levels of development and political power among Maharashtra's subregions.

11. Ambedkar, *Thoughts on Linguistic States*, 22.

12. Maharashtra now has six official administrative divisions. I continue to follow the older tripartite division of the state, which conforms to earlier data on electricity.

13. Banerjee and Iyer, "History, Institutions, and Economic Performance," 1190–213. The research finds that the British territories ruled by *zamindari*, or landlord, continue to have lower economic outcomes than districts that were characterized by *ryotwari*, or individual cultivation. Although the research does not confirm a clear mechanism, the authors speculate that the zamindari districts probably have greater class conflict and therefore higher barriers to collective action than areas that had seen less inequality in earlier decades and centuries. Banerjee and Iyer analyze only the British colonial territories. The areas of Marathwada, which were part of the zamindari system of the Nizam's rule, are most likely comparable to the British territories under zamindari.

14. Srinivas, "Social Structure of a Mysore Village," 18.

15. In recent years, various political organizations have demanded that Marathas be included in the state's list of "Other Backward Classes," accepting social demotion to ensure greater access to educational and economic reservations. Meena Menon, "Nation in a State."

16. Among the earliest works from American academics, see Carter, *Elite Politics in Rural India*; and Carras, *Dynamics of Indian Political Factions*. Carter focuses on the relationship between social stratification and control of political and economic institutions, whereas Carras focuses on the relationships between Maharashtra's political factionalism and material interests. Both expound on the inordinately large role of elite Marathas.

17. Desphande, "Caste as Maratha," 7–32.

18. O'Hanlon, *Caste, Conflict and Ideology*, esp. chap. 2.

19. Lele, *Elite Pluralism and Class Rule*, 55–56. Omvedt quotes from a Maratha organization the following politically prescient plea made in 1931, in advance of the census: "Only out of ignorance do people not call themselves 'Marathas.' Educated Marathas should clearly inform any ignorant Maratha. . . . The days of the rule of wealth have gone and the day of the rule of numbers has come; we hope our educated Maratha society will remember this." Omvedt, *Cultural Revolt in a Colonial Society*, 134.

20. Deshpande, "Caste as Maratha," 27.

21. The political evolution of caste continues still. The bulk of this chapter is concerned with politics from the 1960s through the early 2000s, when the success of Congress in Maharashtra rested on the support of the consolidated Maratha-Kunbi communities. Since then, the field of electoral politics in the state has become much more volatile. Palshikar, Deshpande, and Birmal suggest that some of this volatility is due to the

unbraiding of Maratha and Kunbi political identities, as well as the ability of non-Congress parties to appeal to regional factions of these communities. Palshikar, Deshpande, and Birmal, "Maharashtra Polls," 42–47.

22. Lele, *Elite Pluralism and Class Rule*, 52.

23. "Exhibition of Incandescent Lighting," *Times of India*, December 23, 1882.

24. "Gas and Electric Lighting in India," *Times of India*, March 5, 1887.

25. Tata Hydro-Electric merged with two other Tata subsidiaries also in the power sector to become Tata Power Company in 2000. Tata Power Company, "Pioneers in Power," company brochure.

26. Much of the information about the early years of electricity in Bombay is taken from Bombay Electric Supply and Transport Company, "BEST Undertaking."

27. Rutnagur, *Electricity in India*, 16.

28. Reliance Industries assumed a controlling interest in Bombay Suburban Electric Supply (BSES) in 2002; since 2008, the company has been known as Reliance Infrastructure.

29. Maharashtra State Electricity Board, *Administrative Report 1961–1962* (Bombay: Government Central Press), chap. 5.

30. Maharashtra Economic Development Council, *Power Development in Maharashtra* (Bombay: MEDC, 1975), 25.

31. Ibid., 22.

32. Calculated from data in Maharashtra Economic Development Council, *Power Development in Maharashtra*, 75–76, appendices 29 and 30.

33. Constituent Assembly Debates (CAD), vol. VI, part II, 1948, 55.

34. Chibber, *Locked in Place*, esp. chaps. 4 and 6.

35. Central Water and Power Commission, *Public Electricity Supply: All India Statistics 1959–1960* (New Delhi: Ministry of Irrigation and Power, Government of India, 1961).

36. See Mukherjee and Subramanian, "Corporates That Have Stood the Test of Time." Out of the 197 companies that were listed on the Indian stock exchanges at independence in 1947, seventeen were electricity companies.

37. Central Water and Power Commission, *Public Electricity Supply: All India Statistics 1959–1960*, 15. In 1959–60, all of the private companies in Bombay State together generated 2,994,676 kWh, of which almost all would have been in the cities of Bombay, Surat, and Ahmedabad. Private companies in West Bengal generated 1,997,533 kWh, almost all of which would have been in Calcutta. The total generation in India at that time, including all State Electricity Boards (SEBs), municipal utilities, and private companies, was only 15,032,953 kWh.

38. Committee on Public Undertakings 1979–80 (Fifth Maharashtra Legislative Assembly), "Seventh Report on Maharashtra State Electricity Board Regarding Recent Power Cut Imposed by Government of Maharashtra" (Bombay: Government of Maharashtra, 1979), 15. The reports of the Committee on Public Undertakings constitute a valuable archive to uncover conflicts surrounding the function of the public utility, the relationship between the board and the government, and how the selective distribution of electricity mediated state-society relations. The committee, which is composed of state legislators, is charged with investigating and suggesting remedies for the major problems in the functioning of state-level public-sector enterprises revealed by the annual audits.

39. Other sectoral allocations were: irrigation (22 percent), agriculture (22 percent), cooperative sector (12 percent), transport and communications (12 percent), and industries

(5 percent). Finance Department (Planning), *Third Five-Year Plan Maharashtra State: Annual Plan for 1965–66, Part I: Outline* (Bombay: Government of Maharashtra, 1965).

40. Committee on Public Undertakings 1978–79 (Fifth Maharashtra Legislative Assembly), "Sixth Report" (Bombay, Government of Maharashtra, 1979), 4. During the committee's meetings, the utility representative asked for increases to the Maharashtra State Electricity Board (MSEB) budget, which would have increased the overall share of energy-sector spending to 45 percent of the state's development spending.

41. Finance Department (Planning), *Draft Third Five-Year Plan of the Maharashtra State: 1961–1962 to 1965–66* (Bombay: Government of Maharashtra, 1960), 11.

42. Central Electricity Authority, *All-India Statistics, 1971–72* (New Delhi: Government of India, 1972).

43. Committee on Public Undertakings (Third Maharashtra Legislative Assembly), "Eighth Report on the Problems of Low Voltage and Irregular Supply of Electricity with Special Reference to Its Prevalence in Baramati Area" (Bombay: Government of Maharashtra, Maharashtra Legislature Secretariat, 1970), 2.

44. Ibid., 9–10, annex II.

45. Committee on Public Undertakings 1968–69 (Third Maharashtra Legislative Assembly), "Second Report on Power Cut Imposed by Government on Industrial and Domestic Consumers" (Bombay: Government of Maharashtra, 1969), 1–3.

46. Committee on Public Undertakings 1975–76 (Fourth Maharashtra Legislative Assembly), "Fifteenth Report on Maharashtra State Electricity Board Regarding Recent Power Cut Imposed by the Government of Maharashtra," 1–40. The report analyzes the multiple reasons for electricity shortage from 1972 to 1975, including inadequate rainfall for three successive monsoons from 1972 to 1974, equipment repairs and maintenance, and an inadequate transmission network that makes it difficult for the board to move power from where it is produced to where it is needed. The board representative also complained to the government of delays in getting adequate foreign exchange for parts and equipment from foreign manufacturers. See esp. 5–12.

47. Committee on Public Undertakings 1979–80 (Fifth Maharashtra Legislative Assembly), "Seventh Report on Maharashtra State Electricity Board Regarding Recent Power Cut Imposed by Government of Maharashtra," 20, annex I.

48. Ibid., 1.

49. Maharashtra Economic Development Corporation, *Power Development in Maharashtra*, 75, appendix 33.

50. Godbole, *Unfinished Innings*, 128.

51. Author's interview with Madhav Godbole on July 31, 2003, at his home in Pune, India.

52. Sant and Dixit, "Beneficiaries of the IPS Subsidy," 3315–21. The study is necessarily deductive since the MSEB's data about agricultural consumption are not reliable, a quality not unique to MSEB.

53. EPW Research Foundation, *District-Wise Agricultural Database for Maharashtra*, annex II: "Area Under Different Crops: Maharashtra."

54. Ibid., annex IV.

55. *Human Development Report Maharashtra 2002* (Mumbai: Government of Maharashtra, 2002), 36–37. The report was collaboratively produced by the United Nations Development Program, the Planning Commission, and the government of Maharashtra. The involvement of various government agencies does not preclude the report's authors

from directing serious criticism at the functioning of government, as the quoted text illustrates.

56. Manu Goswami highlights the dramatic increases in investments flowing into the Indian subcontinent after the British crown displaced the British East India Company in 1857. She argues that this can be explained by considering that the British relied on "state works"—principally expanding rail networks and canal systems—as the physical markers that signified the transition from the mercantile rule of the company to formal colonial rule. Goswami, *Producing India*, 42–72.

57. Atwood, "Peasants Versus Capitalists," 59–80.

58. Mann and Kanitkar, *Land and Labour*; and Inglis and Gokhale, "Note on the Outturn and Profit from Sugarcane Crops"; cited in Atwood, "Peasants Versus Capitalists," 66.

59. Attwood, *Raising Cane*, 50–67.

60. Tanks are commonly used throughout peninsular India. Most often, a tank is maintained in a naturally occurring dip in the land where rainwater collects or by blocking the flow of water from a stream with an earthen embankment. For more on premodern irrigation systems, see Williamson, "Indigenous Irrigation Works," 613–26.

61. Mitra and Muranjan, *Economic Benefits of Rural Electrification*, 57. Additional reasons that oil engine owners gave for not converting to electric once it became available were the inadequacy of water in the well or lands that could be easily irrigated from the well site, and MSEB delays in providing connections.

62. *Report of the Sugar Industry Enquiry Commission: 1974, Volumes I and II* (New Delhi: Ministry of Agriculture, Government of India, 1974), 77.

63. Baru, *Political Economy of Indian Sugar*, 84.

64. Amin, *Sugarcane and Sugar in Gorakhpur*, chap. 7. Cited in Baru, *Political Economy of Indian Sugar*, 33.

65. Kamat, *Management of Co-operative Sugar Factories*, 38. Kamat's study of thirteen cooperative factories in 1961 found that while farmer members of some of the cooperatives relied almost wholly on canal waters, members of other cooperatives that had farmed in canal tracts nevertheless used a mix of well and canal waters, and in still others, wells were used almost exclusively. Cane-farming members of the Pravara Sugar Factory, the first cooperative factory to be set up in Maharashtra following independence, cultivated a total of 4,634 acres of sugarcane. Out of this, despite the factory's proximity to the Pravara Right and Left Bank Canals, farmers irrigated 85 percent of the land with well water and only 15 percent with canal waters.

66. In 1932, the colonial government of India had raised tariffs on imported sugar (most of which was coming from Java), which gave a fillip to local sugar production.

67. Attwood, *Raising Cane*, 88–89.

68. D. R. Gadgil was an important economist, not just in the state but in all of India; he retired in the late 1960s as head of the central planning commission.

69. Attwood, *Raising Cane*, 194.

70. Ibid., 190.

71. Baru, *Political Economy of Indian Sugar*, 86.

72. Manor, "Where Congress Survived," 785–803.

73. Chousalkar, "Co-operatives," 34–35.

74. This could be compared to the way that diverse Indian political parties have relied on India's vibrant, and occasionally violent, student politics in colleges and universities to

yield new political stars. Of the current crop of influential politicians in India, Lalu Prasad Yadav of Bihar, Mulayam Singh Yadav of Uttar Pradesh, and Buddhadeb Bhattacharya from the Communist Party of India (Marxist), or CPI(M), in West Bengal all got their start in student politics. In Maharashtra, many of the most influential politicians of the last thirty years got their start in cooperative politics, including Sharad Pawar, Vasantdada Patil, and Ajit Pawar. The one Maratha who is remembered as having attempted to weaken the dominance of the "sugar barons" in Maharashtra's politics was Shankarrao Chavan. He was a Maratha, but from Marathwada rather than Western Maharashtra, who studied in Hyderabad and practiced law before running for elected office.

75. Khekale, *Pressure Politics in Maharashtra*, 93–95. During interviews conducted by Khekale, two presidents of the Maharashtra Pradesh Congress Committee, the regional party office, confirmed that when deciding which candidates would be given tickets to contest elections, they explicitly looked for experience in running cooperatives as evidence of political aptitude and chances of success.

76. Ibid., 92.

77. Shinde, *Working of the Maharashtra Legislative Assembly*, 34–35.

78. Ibid., 105.

79. Author's interview with Sharad Joshi at his home in the village of Ambethan, Maharashtra, October 9, 2002. In the interview, Joshi stated that since the Shetkari Sanghatana's Sangli *adivesan* (meeting) in 1990, the organization has advised its members to stop paying electricity bills.

80. Lindberg, "New Farmers' Movements in India," 95–125.

81. "Power Tariff for Farms Cut to 20 p." *Times of India*, August 10, 1978.

82. Central Electricity Authority, *Electric Power Survey of India 1987* (New Delhi: Government of India, 1987). In 1984–85, captive power units in Maharashtra generated 692 million kWh of electricity, while the MSEB generated 22,773 million kWh. In Gujarat, there was more captive power generated (786 million kWh), although the Gujarat electricity board was roughly half the size of the MSEB in terms of total capacity and total generation (12,806 million kWh).

83. "MSEB Turns Corner," *Times of India*, March 24, 1987; "Power Deficit in State Feared," *Times of India*, December 15, 1987.

84. Author's interview with P. Abraham in New Delhi, April 2, 2003.

85. "Maharashtra Pushes for Gas-Fired Plants Despite Opposition," *Power Asia—FT Energy Newsletter*, September 25, 1989.

86. "Why Maharashtra SEB Might Be Worth Privatizing," *FT Energy Newsletter—Power Asia*, March 25, 1991.

87. Gulati and Narayanan, *Subsidy Syndrome in Indian Agriculture*, 105. Gulati and Narayanan use central government data to show that the power subsidy is highest in Gujarat, followed by Maharashtra. Although the price is higher per unit in those states than in others, the sheer volume of power consumed by farmers in these western states makes their subsidies the largest.

88. Author's interview with Dilip Walse-Patil at his home in Mumbai, August 8, 2003. Walse-Patil, energy minister at the time, comes from a politically-connected Maratha family and is a close associate of Sharad Pawar, whose photo adorned one wall of his office when we met. Since 2009, Walse-Patil has been speaker of the State Legislative Assembly.

89. "State Lets Private Sector into Power," *Times of India*, September 13, 1990; Debashis Basu, "14 Bids for State Power Projects," *Times of India*, December 14, 1990.

90. In addition to a host of articles in the Indian press, there is extensive coverage of the investment project in *Economic and Political Weekly*. In addition to books and articles on the topic cited earlier, see Ahmed, "Neoliberalism, Corporations, and Power," 621–39.

91. "No Gas-Based Power Plant at Dabhol," *Times of India*, April 26, 1989.

92. Central Electricity Authority, "Power Survey of India—14th" (New Delhi: Government of India, 1991), 72, table showing "Installed Capacity, Peak Availability, Peak Load, Energy Availability and Energy Requirement" in Maharashtra.

93. Mehta, *Power Play*, 18–21.

94. World Bank analysis as quoted in ibid., 38–39.

95. Vidhyadar Date, "Investors Watching State Elections," *Times of India*, January 30, 1995; Ambarish Mishra, "BJP to Focus on Economic Issues," *Times of India*, December 21, 1994.

96. About the swiftness and secrecy with which the project was renegotiated, a later government report notes dryly: "It is a matter of some concern that deliberations that formed the basis for a decision by the GoM to support a project that was potentially a liability for the State to the extent of around Rs. 6000 crore a year (at today's exchange rate) should be recorded in so summary a fashion." *Report of the Energy Review Committee, Part I* (known colloquially, and referred to hereafter, as the "Godbole Committee Report"), 2001, 27. An observation by the Bombay High Court about the renegotiated agreement was even more blunt in its assessment: "The speed at which the whole thing was done by the negotiating group is unprecedented. What would stop one to say, as was said by the Chief Minister in the context of the original PPA, Enron revisited, Enron saw and Enron conquered—much more than it did earlier." As quoted in Godbole Committee Report, 39.

97. Ibid., 29.

98. "Anti-Enron Movement Denounced Govt. Move," *Times of India*, June 4, 1995; Ashley D'Mello, "Residents Set to Oppose Land Acquisition Near Dabhol," *Times of India*, September 10, 1996.

99. Human Rights Watch, *Enron Corporation*.

100. Godbole Committee Report, 3.

101. Ibid., 83–93.

102. "Bechtel Sells Its Stake, Breathing New Life into Dabhol Power as an Indian-Owned Firm." *Global Power Report*, July 21, 2005.

103. Prasad, Prasad, and Reddy, "Bio-Gas Plants," 1347–64.

104. World Bank, *Asia Sustainable and Alternative Energy Program*, 7–8.

105. The precursor institutions to the Ministry of New and Renewable Energy (MNRE) were the Commission for Additional Sources of Energy, established in 1981; the Department of Non-Conventional Energy Sources, formed in 1982; and the Ministry of Non-Conventional Energy Sources (MNES) in 1992. The MNES was rechristened the MNRE in 2006. Within Maharashtra, the state created the Maharashtra Energy Development Agency (MEDA) with the aim of finding decentralized mechanisms for rural electrification. In the 1990s, the MEDA shifted its focus to include setting up decentralized rural systems, but then progressively linking them to the grid.

106. Haya, Ranganathan, and Kirpekar, "Barriers to Sugar Mill Cogeneration in India," 67.

107. Ibid., 74.

108. "MERC Order for Non-fossil Fuel Based Co-generation Projects" (Mumbai: Maharashtra Electricity Regulatory Commission, 2002), 1–125, http://mercindia.org.in/pdf/16082002.pdf (accessed May 17, 2012).

109. Ibid., 28, 38.

110. Information available at the Web site of the Maharashtra Energy Development Agency, http://www.mahaurja.com/PDF/PG2_bagase_Projcomm.pdf (accessed May 23, 2012).

Chapter 4

1. For details about the financial and technical performance of the State Electricity Boards (SEBs) during the 1990s, see the *Annual Report on the Working of the State Electricity Boards and Electricity Departments*, produced by the Planning Commission's Power and Energy Division.

2. On the related idea of "homegrown conditionality," see Chaudhry, Kelkar, and Yadav, "The Evolution of 'Homegrown Conditionality' in India," 59–81.

3. See for example Abhijit Banerjee and Somanathan, "The Political Economy of Public Goods," 287–314.

4. Only a few other analyses of Odisha's privatization point out the low levels of electricity consumed by the farm sector. Dubash and Rajan, "Power Politics," 3367–90, suggests that Orissa privatized because the state had no rural opposition and its chief minister believed the state utility would be imminently bankrupt. Thillai Rajan, "Power Sector Reform in Orissa," 657–69, provides a typology of three layers of causal factors—context, trigger, and facilitator—and lists a number of variables in these categories, including World Bank lending terms, OSEB insolvency, state fiscal crisis, support of OSEB chairman, small farm sector, and support from state and central government. Ramana, Mishra, and Nayak, "Power Sector Reform in Orissa," 376–408, points to the poor performance of OSEB as the main cause of the reform program. A government-supported perspective is found in *Report of the Committee on Power Sector Reform of Orissa* (Bhubaneshwar: Government of Orissa, October 2001). The report, known colloquially and hereafter as the "Kanungo Committee Report," named after the chairman, also proposes that the primary reason for the reforms was OSEB's inadequacy, but further suggests that the shape of the program was guided by World Bank plans.

5. Manoranjan Mohanty, "Class, Caste and Dominance," 321–66.

6. Article 46 of the Indian Constitution enjoins the Indian state to provide special protections for the educational and economic interests of the Scheduled Castes (SCs), or former ex-untouchables, and Scheduled Tribes (STs), also called *adivasis*, or indigenous people.

7. Sinha, *Geography of Orissa*, 182.

8. Odisha has 98 percent of the country's chromite reserves, 35 percent of nickel, 23 percent of manganese, 23 percent of dolomite, 24 percent of coal, and 22 percent of iron ore. See Johri, "Nalco's Role in Orissa's Development," 342–46.

9. B. K. Mohanty, "Mineral Development Prospects," 453–58.

10. In 2010–11, exports made up more than 30 percent of the company's total sales: National Aluminum Company Limited, *30th Annual Report, 2010–2011*. Available online

at http://www.nalcoindia.com/investor/30th_Annual_Report.pdf (accessed June 7, 2013).

11. U. K. Mohanty, "Potential for Coal Mining," 459–66.

12. Misra, *Party Politics and Electoral Choice*, 42–44.

13. Jena, "Political Parties in Orissa," 485–502.

14. Anderson, *Imagined Communities*. For a detailed history of the language movement from its origins in the late nineteenth century through the constitution of British Orissa in 1936, and including the integration of princely states after independence, see Patra, *Formation of the Province of Orissa*.

15. This was part of the Congress resolution adopted in 1920 in Nagpur and laid the groundwork for the linguistic federalism that followed independence, in which new states were created to unite linguistic communities.

16. Manoranjan Mohanty, "Class, Caste and Dominance," 320. The British inherited the system of treating the coastal and inland areas in two distinct manners from the Mughal emperors. However, whereas under the Mughals, the princes were treated as *zamindars* who had the right to collect rent in an estate, the British vested them with proprietorship, which changed the relations between the princes and the villagers in their territories. For a discussion of how British policy transformed rural land relations, see also Rath, *Law of Zamindari Abolition*, 19–20. The book by Rath provides the text of the law and commentary by him, and was written as a field manual for law enforcers.

17. On how the rajas cooperated with the British during the noncooperation movement of 1920–22, see Pati, *Resisting Domination*, 62–69.

18. Manoranjan Mohanty, "Class, Caste and Dominance," 348.

19. Jena, "Political Parties in Orissa," 486.

20. Misra, *Party Politics and Electoral Choice*, 47–49.

21. For an ethnographic analysis of how erstwhile royals exploited traditional social relations for electoral gain, as well as many other aspects of Odisha's first decade of electoral politics, see Bailey, *Politics and Social Change*, esp. chap. 4.

22. Banerjee, *Regional Political Parties*, 70–72. See also Bailey, *Politics and Social Change*, 173–77 and 197–206.

23. Banerjee, *Regional Political Parties*, 201, table 7.10.

24. Ibid., 297, 311.

25. Pati, "Dialectics of Transition," 353–64.

26. Erdman, "India's Swatantra Party," 394–410.

27. This subject awaits extensive and dedicated research. Here, I draw from a variety of English-language secondary materials on politics in Odisha, as well as from some early census materials, to piece together a preliminary account of the dynamic relationship between caste and politics.

28. Part of the special protections that the Indian Constitution accords to SC and ST communities are reserved electorates at the national and state levels. The number of constituencies that are reserved, from which only members of either SC or ST communities can stand for election, is proportional to the population of these communities within the state as a whole.

29. Election Commission of India, "Statistical Report on the General Elections to the Legislative Assembly of Orissa" (New Delhi: Government of India, various years).

30. Manoranjan Mohanty, "Class, Caste and Dominance," 347.

31. Chandra, *Why Ethnic Parties Succeed*, see 159 for a ranking of all states according to their level of support for the Bahujan Samaj Party (BSP).

32. Writing about politics in Odisha at the end of the first decade of competitive elections, Bailey notes that whereas caste associations had formed in other parts of India to consolidate smaller *jati* (subcaste) identities into larger political units, he finds no evidence of these tendencies in Odisha. See Bailey, *Politics and Social Change*, chap. 5.

33. The middle, peasant castes are officially referred to as "Other Backward Classes." The 1931 Census, the last census to enumerate along caste lines, records that Khandayats made up roughly 10 percent of the state's population, as cited in Sanjay Kumar, "Janata Regionalized," 111–38, fn. 12.

34. Ibid., 115–16.

35. Subramanian, *Ethnicity and Populist Mobilization*, 94–95.

36. Census of India, 1921. *Volume VII: Bihar and Orissa, Part I* (Patna: Government Printing, Bihar and Orissa, 1923), 233. By the 1920s, although there is no marriage between castes, there is evidence of some marriage among different *jatis*, or subcastes, within a larger caste formation.

37. Census of India, 1931. *Volume VII: Bihar and Orissa, Part I* (Patna: Government Printing, Bihar and Orissa, 1933), 267.

38. Ibid., 267–68. The census reports that efforts by lower caste groups to claim higher social status led to riots by Khandayats and Karans in Cuttack District.

39. Manoranjan Mohanty, "Class, Caste and Dominance," 340.

40. Shah, *In the Shadows of the State*, 18.

41. Menon and Nigam are careful to chart the earlier histories of displacement and dispossession, but suggest that these processes have accelerated in the current period and are now in the service of private corporations rather than the state-led development projects. Menon and Nigam, *Power and Contestation*, 61–82.

42. Nayak, "Big Dams and Protests in India," 69–73. For a larger study of opposition to big dams in India, see Khagram, *Dams and Development*.

43. This chapter focuses on the postcolonial period, but biographies of preindependence political leaders like Madhusadan Das and Gopabandhu Das suggest that extractive industrialization was an important plank in early nationalism as well. There is also some evidence that the natural resource wealth in the upland areas was part of the rationale for Congress leaders to push for the incorporation of British Orissa with the princely states in a united state; without that mineral wealth, these regional leaders in the Congress Party doubted the economic viability of their state. See, for example, Barik, "Gopabandhu and the National Movement in Orissa," 40–52.

44. Author's interview with Birendra Nayak, Utkal University, Bhubaneshwar, August 8, 2009. Nayak is a professor of mathematics at Utkal University, and is an authority on the state's political history and social movements, or people's movements.

45. Debashish Munshi, "Orissa Steel Project Burdens Patnaik," *Times of India*, July 12, 1990.

46. Baru, "Economic Policy," 207–30. Baru's essay surveys regional capitalist classes in a handful of southern and western states. See also Upadhya, "Farmer-Capitalists," 1376–82 and 1433–42; Chari, *Fraternal Capital*.

47. Debdas Banerjee and Ghosh, "Indian Planning," 104–65.

48. Sinha, *Regional Roots of Developmental Politics*.

49. Davis, *Discipline and Development*.

50. See, for example, Ahmad, "Regional Pattern of Industrial Licensing," 3–17; and Biswas and Marjit, "Political Lobbying and Fiscal Federalism," 716–25. Biswas and Marjit look at how politics impact the way that industrial licenses were issued rather than industrial investment. They also cite a string of government reports that question the effectiveness of the government's policy to promote balanced regional growth through public investments, including the Hazari Report (1967), the Dutt Committee (1969), and the Dagli Committee Report (1979).

51. Responding to this inequity, at one point in his political career when his own allies were in power at the center, Odisha's most storied politician, Biju Patnaik, theatrically threatened secession: Kanungo, "Biju Patnaik," 59–75.

52. A related literature in fiscal federalism looks at the relationship between political leverage and financial transfers from the central government. For an analysis based on data from 1983 to 1992, see Singh and Vasishtha, "Patterns in Centre-State Fiscal Transfers," 4897–903.

53. "Taming Rivers," *Times of India*, May 10, 1947.

54. "Power Supply from Hirakud: Board Appointed to Fix Rates," *Times of India*, May 11, 1953; "The Hirakud Dam: Expected Returns," *Times of India*, June 16, 1953.

55. Central Water and Power Commission (Power Wing), *Public Electricity Supply: 1959–60*, table XLI, "Segregation of Sales in States," 56–57.

56. Programme Evaluation Organisation, *Report on Evaluation of the Rural Electrification Programme*, 18.

57. Irrigation and Power Department, Orissa, *Report on Tikerpara Dam Project, 1964* (Bhubaneswar: Government of Orissa, 1964), i.

58. Khosla, *Orissa's Decade of Destiny*, 9.

59. Planning Department, Orissa, *Third Five-Year Plan, Orissa, Annual Plan 1965–66* (Bhubaneswar: Government of Orissa, 1965), 1.

60. The Khanna Commission was created under the Commissions of Inquiry Act, 1952, which authorizes either a state or the central government to appoint a panel, often staffed or chaired by acting or retired judges, to investigate any "matter of public importance." See Orissa (India) and H. R. Khanna, *Report of the Commission of Inquiry* (Bhubaneswar: Home Department, 1969). From 1959 to 2000, a total of five chief ministers of Orissa were investigated by such commissions on the grounds that they used their public office to amass private wealth. Digambar Mishra, *Political Behavior of Indian State Governors*, 46–47. Mishra makes the argument that these charges of corruption were a function of the political instability of state politics and central to the declaration of several constitutional emergencies throughout this period, the longest one lasting for more than a year in 1973.

61. *Khanna Commission Report*, 811.

62. The last two occasions were coterminous with Indira Gandhi's "Emergency" and then the victory of the Janata coalition at the national level, so they had as much to do with politics in New Delhi as in Bhubaneswar. Digambar Mishra, *Political Behavior of Indian State Governors*, 79–104.

63. This portion of the chapter draws extensively from Planning and Coordination Department (State Evaluation Organisation), *Report on the Evaluation of Rural Electrification Programme in Orissa* (Bhubaneswar: Government of Orissa, 1982).

64. Ibid., 20.

65. Ibid., 69, appendix 12. The average return per acre of nonirrigated land in the study was Rs. 466, while the return for land with access to electrified pump sets was Rs. 1,408.

66. Ibid., 44.

67. Ibid., 89, appendix 34.

68. Orissa State Electricity Board, "About Us" (Bhubaneswar: Government of Orissa, 1976), 23.

69. Ibid., 51.

70. Department of Power, *Report: 1982–1983* (New Delhi: Ministry of Energy, Government of India, March 1983), cited in Centre for Monitoring the Indian Economy, *Current Energy Scene in India*, 60.

71. "Department of Power, Annual Report 1990–91" (New Delhi: Ministry of Energy, Government of India, 1991), 12.

72. The village electrification data indicate only whether the electricity grid has reached a village; these data do not indicate household electrification, which remains low even in states that claim 100 percent village electrification. In Odisha, there are additional reasons to believe that the official electrification rates are overly optimistic. Tribal populations living in the hills often inhabit several settlements dispersed in a region. Usually, the government enumerates one of these—the one with the largest revenue generation—as the primary settlement. Even when the core settlement is electrified, the ancillary hamlets most likely are not, because of which the electrification rate most likely is still lower. See Meher, "Infrastructure," 54–70.

73. Centre for Monitoring the Indian Economy, *Profiles of States*, 297.

74. The two families have no direct kinship relation, although they are both part of the same community, or *jati*.

75. Manoranjan Mohanty, "Adrift Middle Class," 1445–46.

76. "Orissa Power Cuts Will Hit Units," *Times of India*, June 4, 1982; "Industry Blames Orissa Govt.," *Times of India*, July 7, 1984; "Steps to Tide Over Orissa Power Crisis," *Times of India*, December 20, 1987; Debmuni Gupta, "Power Crisis Hasn't Crippled Industry: Patnaik," *Times of India*, January 15, 1988.

77. D. C. Das, "Biju Patnaik," 54–58. The fact that Patnaik headed these two ministries suggests how important Orissa's mining and mineral resources were in Patnaik's vision of development.

78. Manoranjan Mohanty, "Class, Caste and Dominance," 356.

79. Ruben Banerjee, "Iron Hand: Biju Patnaik Asserts Himself," *India Today*, April 15, 1990, 45, 48.

80. Ruben Banerjee, "Clean Sweep: Biju's Charisma Carries the Day," *India Today*, March 31, 1990, 57, 59.

81. Election Commission of India, "Statistical Report on General Election, 1990 to the Legislative Assembly of Orissa" (New Delhi: Election Commission, 1990).

82. Banerjee, "Clean Sweep."

83. "New Steel Plant for India," *Times of India*, April 27, 1990.

84. Biju Patnaik, remarks made to the Meeting of the National Development Council, June 18–19, 1990.

85. Padhi, "Economic Growth," 71–106. The data on the sectoral composition of the GDP is from 76, table 3.4.

86. The agricultural sector has continued to decline into the present, whereas the relative sectoral importance of mining has increased tremendously. For a detailed intersectoral and interdistrict analysis of Odisha's economy during the past twenty years, see Banikanta Mishra, "Agriculture, Industry and Mining," 49–68.

87. Centre for Monitoring the Indian Economy, *Basic Statistics*, cited in Keshab Das, "Industrialisation in Orissa," 200–222.

88. Biju Patnaik, remarks made to the Meeting of the National Development Council, June 18–19, 1990.

89. Biju Patnaik, interviewed by Sabina Seghal, *Times of India*, June 13, 1993.

90. Biju Patnaik, "Soft State, Hard Decisions," *Times of India*, June 13, 1993, as told to Sabina Seghal; reprinted in "Brainstorming Deliberations of Biju Patnaik," *Orissa Review* (February–March 2005), 42.

91. Author's interview with S. K. Mohapatra at his home in Bhubaneswar, August 5, 2009.

92. For number and size of Industrial Entrepreneurs Memorandums (IEMs), see Ministry of Commerce and Industry, Government of India, *Annual Report 2001–2002.*

93. For an excellent comparison of spatial patterns in industrial investment before and after the start of market reforms, see Chakravorty, "Industrial Location in Post-reform India," 120–52.

94. The first steel plant in Rourkela, Orissa, was also the first public-sector steel plant to be built by the Indian state under the company name Hindustan Steel Private Limited, which was later changed to the current name, Steel Authority of India Limited (SAIL). Biju Patnaik promised that he would build a second steel plant for the state during his election campaign.

95. Kunal Bose and Reuter, "Indian Steel Plant Planned," *Financial Times*, January 5, 1993.

96. Kunal Bose, "Indian Alumina Export Project Planned," *Financial Times*, October 15, 1992.

97. Kunal Bose, "Indian Companies Link to Build Alumina Plant," *Financial Times*, November 3, 1992.

98. Kunal Bose, "Indians Press for Ban on Chrome Ore Export," *Financial Times*, May 10, 1991.

99. "Commodities at Risk; Oil's Poor Relations," *Economist*, September 1, 1990.

100. Kunal Bose, "India Aims to Boost Granite Exports," *Financial Times*, August 31, 1990.

101. This, in effect, would realize the vision of Ambedkar and others during the 1940s, of an integrated national grid in which the country's various resources of water and fuel could be shared from one region to the next. What technology was making feasible through the advancing national grid, however, was still obstructed by the array of state-level and regional institutions and interests that had come to fill the energy sector.

102. Central Electricity Authority, *Report of the Committee to Recommend Next Higher Size of Coal Fired Thermal Power Stations.* This report references the findings of an earlier committee, the Advisory Group on Technology Development, housed within the Ministry of Irrigation and Power. This earlier committee—which submitted an initial report in April 1986, was reconstituted in 1989, and completed a second report in April 1990—advocated increasing the average size of new power plants to 750 MW.

103. "South Indian Coal-Fired Schemes Cleared by Delhi," *FT Energy Newsletters—Power Asia*, January 14, 1991.

104. International Coal Report, "India Plans to Export to France," *FT Energy Newsletters—International Coal Report*, June 26, 1992.

105. "US Group Considers Big Coal BOT Scheme in Ib Valley," *FT Energy Newsletters—Power Asia*, December 16, 1991.

106. "US Group Wins BOO Deal in Orissa State," *FT Energy Newsletters—Power Asia*, February 10, 1992.

107. Ibid.

108. "More US Groups Join Privatisation Jamboree," *FT Energy Newsletters—Power Asia*, July 13, 1992.

109. "AES Supplants So. Electric in Ib Valley Project," *FT Energy Newsletters—Power Asia*, January 18, 1993.

110. "Orissa Fields Draw Bucks from US Groups," *FT Energy Newsletters—Power Asia*, October 25, 1993.

111. "Siemens Bids to Build Up Stakes in Power Sector," *FT Energy Newsletters—Power Asia*, March 9, 1992.

112. "Orissa Yields on Private Sector Payment Guarantees," *FT Energy Newsletters—Power Asia*, November 30, 1992. Odisha was the first to provide such a guarantee, although the practice became more common in later years.

113. "AES Supplants So. Electric."

114. "Financial Close Nearly Set for AES Ib Valley Scheme," *FT Energy Newsletters—Power Asia*, September 19, 1994. Power purchase agreements (PPAs) were routinely negotiated by multinational power companies, particularly for their projects in developing countries. They are useful as a way of allocating risk; in the case of the deals signed by Enron in Maharashtra and AES in Odisha, some argue that the risk was shifted entirely to the governments, thereby jeopardizing the longer-term prospects of the projects.

115. Ibid. The financing structure allowed a debt-to-equity ratio of 3.2 to 1, higher than the norm at the time. The equity contribution between the Orissa Power Generation Corporation (OPGC) and AES Transpower was $50 million and $80 million, respectively. Furthermore, although the US Exim Bank was providing a loan to cover 85 percent of the cost of US goods and services, four Indian public-sector banks provided a guarantee for this loan, which shifted the risk to Indian institutions.

116. "DESU Slams in Rate Rises," *FT Energy Newsletters—Power Asia*, September 13, 1993.

117. Power Asia, "Orissa Tries Privatization," *FT Energy Newsletters—Power Asia*, November 22, 1993.

118. "OSEB Gets Set to Restructure," *FT Energy Newsletters—Power Asia*, October 31, 1994.

119. Energy and Infrastructure Operations Division, Country Department II, South Asia Region, *Staff Appraisal Report, India: Orissa Power Sector Restructuring Project* (New Delhi: World Bank, 1996), 12–13 (hereafter referred to as World Bank, "SAR: Orissa"). Much of the narrative about Orissa's reform program is drawn from this lengthy report and the *Kanungo Committee Report*. The World Bank's document sets out the aims of the project and how it is to be operationalized, while the latter is an assessment of the program two years after privatization was completed.

120. World Bank Operations and Evaluation Department, "Reforming India's Energy Sector," 2.

121. World Bank Operations and Evaluation Department, "India: The Challenges of Development," 63–68.

122. Orissa Electricity Reform Act, 1995 (Bhubaneshwar: Government of Orissa, 1995). The act was passed on November 28, 1995, received presidential assent in January 1996, and came into effect in April 1996.

123. Quote cited in "Power House in Order," *Outlook*, September 18, 1996.

124. "NTPC Appetite for Troubled Units," *FT Energy Newsletters—Power Asia*, May 30, 1994.

125. "NTPC Complete Take Over of Talchar Units," *FT Energy Newsletters—Power Asia*, June 12, 1995.

126. Balaji C. Mouli, "India: Package to Rescue Orissa Power Reforms Under Way," *Financial Times*, November 10, 1999.

127. World Bank, "SAR: Orissa."

128. *Kanungo Committee Report*, sec. 3.5.

129. "BSES b'thru in Orissa," *FT Energy Newsletters—Power Asia*, October 30, 1995.

130. Because of decades of state monopoly in the electricity sector, there were only a handful of private companies with sufficient expertise or interest in entering the distribution business. A number of other states also were considering offering urban distribution zones to private companies under limited-term management contracts. At the forefront of these efforts was Andhra Pradesh. See Soutik Biswas and Sunil Dasgupta, "Switching to the Private Sector," *India Today*, April 15, 1995, 110–11.

131. "Distribution Deal for BSES," *FT Energy Newsletters—Global Private Power*, September 30, 1996.

132. "Orissa Power Privatisation Runs into Trouble," *FT Energy Newsletters—Power Asia*, May 27, 1997.

133. *Kanungo Committee Report*, sec. 3.8.

134. "India's Orissa Launches Sale of Distribution Assets," *FT Energy Newsletters—Global Private Power*, December 1, 1997.

135. "Orissa Receives Distribution System Submissions," *FT Energy Newsletters—Global Private Power*, April 1, 1998.

136. For a list of the interested firms and their offer prices, see Ramanathan and Hasan, 37. The authors provide a detailed analysis of each stage of the privatization process.

137. "Orissa Distribution Schemes Find Few Bidders," *FT Energy Newsletters—Global Private Power*, January 1, 1999.

138. "BSES Buys Orissa Power Firms," *FT Energy Newsletters—Power in Asia*, April 19, 1999.

139. In 1998, 49 percent equity and management control in Odisha Power Generating Corporation (OPGC) was sold to the American firm AES, which was also in long-standing negotiations with the state to build additional coal-fired power plants at the same site where OPGC operated, in Ib Valley. From the time the thermal plants of OPGC became operational, the company faced persistent problems in recovering bills from the state-owned company. In order for the sale of OPGC to be acceptable to AES, given the prior record of the Grid Corporation of Odisha (GRIDCO), the state had to provide an escrow account guaranteeing that revenues collected from the central region would go directly to the power generator. *Kanungo Committee Report*, sec. 5.32.

140. "Orissa Advances Distribution Company Sales," *FT Energy Newsletters—Global Private Power*, May 1, 1999.

141. *Kanungo Committee Report*, sec. 4.21.

142. "Orissa Power Setback Has a Long History," *Statesman*, August 4, 2001.

143. "Gridco No to AES Offer to Back from Distribution Biz," *Economic Times*, August 4, 2001.

144. Government offices and public-sector units combined accumulated Rs. 144.8 crores in unpaid electricity bills from September 1999 to March 2001. *Kanungo Committee Report*, sec. 4.29.

145. "OERC Blames BSES, AES for Orissa Power Crisis," *Financial Express*, June 28, 2002.

146. Cited in Nikhil Mookerjee, "Why Tomorrow Belongs to Kalinga," *Outlook*, December 29, 1997.

147. Balram Mohanty, *Orissa*, 52, provides the percent of villages electrified for all thirty districts. In 1994–95, the highest electrification rates are found in Cuttack (95.0), Khurda (92.7), Puri (94.1) in the east, and the lowest rates in Rayagada (35.1), Malkangiri (32.2), Gajapati (45.9), and Deogarh (30.4) in the south and west.

148. *Kanungo Committee Report*, sec. 5.54.

149. "Orissa Tariffs Boosted 15%," *FT Energy Newsletters—Power Asia*, August 8, 1994.

150. *Kanungo Committee Report*, annex 8.

151. World Bank, "SAR: Orissa," 30.

152. Ibid., 32.

Chapter 5

1. Reddy, "Politics of Accommodation," 265–321. Reddy's discussion of the eclipse of the Andhra Communists in the 1950s can be found on 279–81. For a lengthier discussion of the social bases, political ideology, and organizational tactics of the Andhra Communists, see Harrison, "Caste and the Andhra Communists," 378–404.

2. Kohli, "The NTR Phenomenon," 991–1017.

3. Andhra Pradesh State Electricity Board, *Administration Report, 1959–60* (Hyderabad: Government of Andhra Pradesh, 1960), 1–3.

4. Andhra Pradesh State Electricity Board, *Third Administration Report, 1961–62* (Hyderabad: Government of Andhra Pradesh, 1962), 10.

5. Planning Department, *Economic Development of Andhra Pradesh, 1951–1968* (Hyderabad: Government of Andhra Pradesh, n.d.); the analysis of the overall plan allocations across budget heads is given on 22–24, with a full accounting of the plans given in annex I on 71–73.

6. Finance and Planning Department, *Sixth Five-Year Plan, 1980–1985, Draft Vol. I* (Hyderabad: Government of Andhra Pradesh, 1980); see 78–79 for an analysis of plan spending from 1974 to 1980.

7. Andhra Pradesh Department of Industries and Commerce, *Industrial Opportunities in Andhra Pradesh* (Hyderabad: Government of Andhra Pradesh, 1964), 7–10. A specially formed committee would determine the precise concession offered to a new industrial venture, but the discount would range between 25 and 50 percent of the existing rates for industrial consumers.

8. Bureau of Economics and Statistics, *Five Decades of Andhra Pradesh* (Hyderabad: Government of Andhra Pradesh, 2006), table 7: "Area Under Principal Crops in Andhra Pradesh from 1956–57 to 2005–06."

9. Kobo, "Cropping Pattern Changes," 209–10.

10. Andhra Pradesh State Electricity Board, *Administration Report 1959–60*, 19–28.

11. Andhra Pradesh State Electricity Board, *Administration Report 1967–68*, 11–12.

12. The shift to a flat-rate system was initially resisted by the utility chairman, who predicted the adverse consequences for the utility's finances. Author's interview with E. A. S. Sarma, principal, Administrative Staff College of India, Hyderabad, March 5, 2003. Sarma worked for the Andhra Pradesh State Electricity Board (APSEB) from 1976 to 1978, when the government was first contemplating switching the billing systems. Due to resistance from him and the then chairman of the utility, N. Tata Rao, the APSEB delayed implementing flat-rate meters until 1982.

13. Transmission Corporation of Andhra Pradesh Limited, *Power Development in Andhra Pradesh (Statistics) 1999–2000* (Hyderabad: AP Transco, 2000), table 9.11.

14. Bentall and Corbridge, "Urban-Rural Relations," 27–48.

15. Durgaiah, *Farmers Movements in India*, 46–50.

16. Ibid., 51.

17. Except for three years from 1966–67 to 1968–69, throughout the 1950s and 1960s, the APSEB provided region-wise data on village electrification that show that coastal districts had much higher rates of electrification than those in Telengana. See Andhra Pradesh State Electricity Board, *Administration Report*, various years.

18. "Bigger Power Cut," *Indian Economic Diary,* April 2–8, VI.14, 3121.

19. Transmission Corporation of Andhra Pradesh Limited, *Power Development in Andhra Pradesh (Statistics) 1999–2000*, table 7.14: "Power Restrictions (1977–2000)."

20. "Supply Cuts Again for Andhra Pradesh Industry," *FT Energy Newsletters—Power Asia*, January 28, 1991.

21. For example, the "Information and Communications Technology (ICT) Policy of the Government of Andhra Pradesh 2005–2010" (Hyderabad: Government of Andhra Pradesh, 2005) allowed small and medium-sized firms a 25 percent rebate on the power bills for three years. The "Andhra Pradesh Policy on Information Technology Enabled Services (ITES) 2002" (Hyderabad: Government of Andhra Pradesh, 2002) exempts ITES firms from the statutory cuts that had crippled industry in the state for the preceding thirty years.

22. The Andhra Pradesh Electricity Regulatory Commission (APERC) announced in October 2003 that no consumer would require consent or permission to install a captive power plant or generator of any size or capacity. APERC Press Release, December 2003.

23. Joseph, "Politics of Power."

24. The Bank's project appraisal document states that "the Government of Andhra Pradesh has launched an ambitious program for restructuring and reforming its power sector, complex and challenging by itself but made even more ambitious by the interlinked objective to use power sector reform as a critical element of structural and fiscal reform in Andhra Pradesh." Energy Sector Unit, South Asia Region, "Andhra Pradesh Power Sector Restructuring Program," Report No. 18849IN (New Delhi: World Bank, January 1999), 8. Naidu, in a public forum in 2002, stated, "I spend maximum time on the power sector. These reforms are important to my political career." "Power Reforms Crucial for My Future, Says Naidu," *Times of India*, April 21, 2002.

25. Country Operations, Industry and Finance Division, South Asia Region, "Andhra Pradesh: Agenda for Economic Reforms" (New Delhi: World Bank, January 1997), 31.

26. Ibid., 36.

27. Deva Kesava Rao, "India CM Completes 2 Years in Office; Sets 'Dream' Agenda," *Hindu*, September 1, 1997.

28. Kirk, *India and the World Bank*. Kirk offers an account of how Andhra Pradesh came to be chosen along with Uttar Pradesh and Karnataka for the Bank's subnational lending program, in which he suggests that both the central government and the World Bank were keenly aware of Naidu's political power vis-à-vis the fragile United Front coalition government that ruled the center from 1995 to 1997, and in which Naidu's Telugu Desam Party (TDP) was a constituent, as well as the Bharatiya Janata Party–led National Democratic Alliance that ruled from 1998 to 2004, in which Naidu's party supported from the outside. See chap. 2.

29. "India Bill Introduced as Power Staff Go on Strike," *Hindu*, April 28, 1998.

30. Ibid.

31. "India Power Bill Passed; Opposition Suspended," *Hindu*, April 29, 1998.

32. *Andhra Pradesh Electricity Reform Act, 1998* (Hyderabad, Andhra Pradesh, 1998). Part VI of the act deals with granting licenses to nonstate actors for transmission and supply (or distribution).

33. Amarnath K. Menon, "Andhra Pradesh: Ramming Home Reforms," *India Today*, May 11, 1998, 30.

34. "Power Tariff Hiked for HT, Bulk Consumers," *Hindu*, December 29, 1998.

35. "Congress Dharna Against Power Rate Hike," *Hindu*, January 6, 1999.

36. The Federation of Andhra Pradesh Chamber of Industries met in February 1999 to publicly urge Naidu to reconsider the policy change, claiming that smaller industries in particular would be pushed to the brink, and even threatening to launch an industrialists' hunger strike. "FAPI Wants Government to Withdraw Power Tariff Hike," *Business Line*, February 15, 1999.

37. "Andhra Pradesh Move to Cut Rabi Paddy Area," *Business Line*, October 12, 1998.

38. Meetings of the leaders of several farm organizations, such as the Peddireddy Thimmareddy Farm Foundation, which had led the earlier agitation against the tariff increase in West Godavari district, and the Rythanga Samakya, were held in October 1997. The leaders voiced their opposition to privatization, particularly in the face of the erratic and poor quality of power supplied by the utility. "Farmers Oppose Bid to Restructure Andhra Pradesh State Electricity Board," *Business Line*, October 7, 1997.

39. An India media report on the reasons for the tariff increases reported that a 10 percent increase in tariffs was part of the World Bank's stipulation for the lending program. "Naidu's Power Problems," *Hindu*, January 3, 1999. However, this does not appear as an explicit conditionality in the World Bank's *Andhra Pradesh Power Sector Restructuring Program* report document.

40. Manor, "Explaining Political Trajectories," 255–84.

41. Ibid., 264. Manor further adds: "Naidu took a few modest steps down this road [of liberalization]—and made much of them to donors and investors." He later points out that Naidu used the bulk of the funds from the World Bank to finance highly populist development policies that were then "advertised as benefits flowing directly from Naidu himself," 265.

42. Just before the election, one account stated that "if he is elected, that will show it is possible to stress productivity and decentralisation, rather than subsidies, and still win. What is at stake is not just Mr. Naidu's future but a change in the prevailing politi-

cal thinking among India's states." "Booting Up in Andhra Pradesh," *Economist*, September 11, 1999; After the election, another account reported that "something amazing happened in Indian politics. . . . In the large southern state of Andhra Pradesh, the party that promised the biggest freebies lost. And the politician who warned that there's no free lunch won." Dugger, "Surprise in Indian State."

43. Harshe and Srinivas, "Vote for Development," 3103–5.

44. "Power Tariff Hike on the Cards," *Hindu*, November 19, 1999.

45. "Riding Two Horses," *Hindu*, June 4, 2000.

46. R. J. Rajendra Prasad, "Opposition Plans Agitation Against Power Tariff Hike in Andhra Pradesh," *Hindu*, May 29, 2000.

47. "Naidu Promises More Subsidy for APTransco," *Economic Times*, June 6, 2000.

48. R. J. Rajendra Prasad, "Power Tariff Hike: Pressure Mounts for a Rollback," *Hindu*, June 14, 2000.

49. "Stalemate Continues in Assembly," *Hindu*, August 17, 2000.

50. "Congress, Left, MIM Members Suspended," *Hindu*, August 19, 2000; "Andhra's Power Play," *Hindu*, August 30, 2000.

51. Alan Beattie, "World Bank Forced to Defend Plans in India," *Financial Times*, November 15, 2000.

52. "AICC Clears Free-Power Campaign," *Times of India*, July 23, 2002; "Give Top Priority to Farm Sector: Naidu," *Times of India*, October 3, 2002.

53. Energy and Infrastructure Unit, South Asia Region, "Implementation Completion Report on a Loan in the Amount of US$210 million to the Government of India for Andhra Pradesh Power Sector Restructuring Project" (Washington, DC: World Bank, February 20, 2004), 23. For an analysis of subsequent phases of power-sector reforms and restructuring in Andhra Pradesh, see Rahul Mukherji, "Globalization and Deregulation," chap. 6.

Chapter 6

1. There is a substantial difference in electrification between rural and urban households in West Bengal, as in other states. Of the rural population, 40.3 percent rely on electricity, while 85.1 percent of urban residents use electric lighting.

2. For a more detailed consideration of politics in Gujarat, which also includes a discussion of the events of 2003–4, when the leader of the farmer's union went on a hunger strike to protest tariff increases, see: Jaffrelot, *Saffron Modernity in India*.

3. For an analysis of Gujarat's economic policies during Modi's tenure as chief minister, including the government's power sector reforms, see Aseema Sinha, "Reforming Public Services in a High Growth State," 151–56.

4. After partition, 33 percent of total irrigated lands, 20 percent of canal-irrigated lands, and 38 percent of well-irrigated lands were in Indian Punjab. Chadha, *State and Rural Economic Transformation*, 27.

5. Ibid.

6. Ibid., 28–29.

7. Canals accounted for 59 percent of net irrigated area in 1950. By 1975, electric tube wells were working on 58 percent of the total irrigated area. Ibid., 92, appendix 3.3.

8. According to the Indian Ministry of Irrigation's Central Ground Water Board, out of Punjab's 138 blocks, 103 are considered overexploited, five are critical, and another

four are considered to be semicritical. Data available online at http://cgwb.gov.in/gw_profiles/st_punjab.html (accessed April 20, 2013).

9. Gupta, "Punjab State Electricity Board," 261–66.

10. One protracted conflict between farmers and the Punjab State Electricity Board occurred in Gurdaspur District in the early 1970s, along the Pakistan border. As a multiclass alliance of small and larger farmers, there were sharp internal differences in the movement's demands. While those with larger landholdings wanted a reduction in the flat-rate tariff, smaller farmers actually wanted to switch to metered power, which would allow them to pay only for the modest amounts of electricity that they consumed rather than the substantially higher minimum flat-rate tariffs. For an analysis of this conflict, see Azad, "Recent Farmers' Agitations in Punjab," 702–6.

11. Jain, "Political Economy of the Electricity Subsidy," 4072–80.

12. Chadha, *State and Rural Economic Transformation*, 28: "In the post-Partition period, in Indian Punjab, there was a steady shift of [political] power from the urban bourgeoisie to the rural elite and the farming population." For a sociological analysis of state legislators in Punjab, see Kumar and Sharma, "Legislative Elite in Punjab," 151–63. Kumar and Sharma show that from 1977 to 2002, the legislature has been dominated by farmers, mostly coming from the "Intermediate caste" of Jat Sikhs.

13. Sawhney, "Subnational Reforms," 50–52.

14. Punjab dismantled its State Electricity Board (SEB) only in 2010, more than ten years later than many other states and seven years after the 2003 national electricity legislation made it mandatory for all states.

15. By "colonial Madras," I refer to the entire region of the Madras Province rather than just the city of Madras/Chennai.

16. The other states with high rates of village electrification were Karnataka (12.2 percent), Kerala (36.6 percent), Andhra Pradesh (13.1 percent), and Punjab (21.9 percent). Of these four, the first three are located in the south and at least some of the districts of these states were part of colonial Madras. The arguments in this section about Tamil Nadu, then, also help us to understand processes in other parts of South India. See Chapter 2 for the full state-wise data on village electrification in 1963, as well as agricultural consumption of electricity.

17. Programme Evaluation Organisation, Planning Commission, *Report on the Evaluation of the Rural Electrification Programme*, 1965, 28, 38–39.

18. Ibid., 42.

19. "98% of Harijan Colonies in TN Electrified," *Times of India*, October 11, 1976.

20. Programme Evaluation Organisation, Planning Commission, *Report on the Evaluation of the Rural Electrification Programme*, 41–42.

21. Ibid., 39.

22. Subramanian, *Ethnicity and Populist Mobilization*, 209.

23. Madras State Electricity Board, "Administration Report of the Madras State Electricity Board, 1961–1962" (Madras: Government of Madras, 1962), 32. Other measures that benefited agricultural consumers included free service connections, lower tariffs than other low-tension consumers, low tariffs for rural agro-industries, and concessional loans for pump sets.

24. The percentage for the *Second Five-Year Plan* includes transfers from centrally sponsored schemes; excluding these funds, the power sector still took up 35 percent of

the state plan. Planning Department, *Third Five-Year Plan, Madras State: Draft Outline* (Madras: Government of Madras, 1960), 1–5.

25. Washbrook, *Emergence of Provincial Politics*, 50–51.

26. Ibid., 48n100.

27. Of all the provinces, Madras was the first to pass an industrial finance act, the State Aid to Industries Act in 1923, following an example set by Princely Mysore a decade earlier. See Samant and Mulky, *Organisation and Finance of Industries in India*, 250–57.

28. Washbrook, *Emergence of Provincial Politics*, 62.

29. Yenda Srinivasa Rao, "Electricity, Politics and Regional Economic Imbalance," 60.

30. "Cheap Electricity for Rural Areas: Justice Party's Plans," *Times of India*, August 3, 1935.

31. "Electricity in Madras: Recent Progress, Govt. and the Public Co-operate," *Times of India*, August 22, 1924; "Electrical Work in India: Recent Progress in Madras," *Times of India*, July 15, 1926; "Electric Power in Madras: Presidency Developments," *Times of India*, March 14, 1929.

32. "Hydro-electrics in Madras: Question of Cheap Power Supply," *Times of India*, March 27, 1930; "Electricity in South India: Growing Demand for Power, Madras Government's Far-sighted Policy," *Times of India*, September 25, 1930.

33. "Hydro-electrics in South India: A National Scheme," *Times of India*, December 4, 1930.

34. More than a decade before the 1948 act mandated electricity boards in the states, the Madras government surveyed the desirability of a Madras board in South India to be modeled along the lines of institutions in England. *Report of the Madras Electricity Board Committee, 1935–36, Volume II* (Madras: Government Press, 1936), iv–xii.

35. Yenda Srinivasa Rao, "Electricity, Politics and Regional Economic Imbalance," 65–66.

36. In the Select Committee debates that preceded the 1948 act, Madras representative T. T. Krishnamachari bitterly dissented from amendments that allowed private firms to continue to operate in the sector. He blamed his fellow committee members for their "surrender to the persistent agitation that has been carried out by vested interests in the country for some years past designed to thwart all ideas of nationalisation of industries." As the largest firm in the electricity industry, among the interests to which he referred were likely to be the Tata power companies. "Report of the Select Committee on the Electricity (Supply) Bill, 1947," *Constituent Assembly of India (Legislative)*. The quote comes from the "Minutes of Dissent" written in March 1948. I discuss the 1948 act in more detail in Chapter 2.

37. From 1951–52 to 1975–76, the production of sugarcane increased more than fourfold, and that of paddy nearly tripled. Kurien, "Dynamics of Rural Transformation," 373–74.

38. "The Politics of Power," *Economic and Political Weekly* 8, no. 15 (April 14, 1973): 700–701.

39. On the growth of all subsidies in Tamil Nadu, including those for electricity, see Ambirajan, "State Government Subsidies," 811–21. The gradual increase in subsidies was due not just to government initiatives but was also actively demanded by farmers' movements, as has happened in other states. See Rajagopal and Anbazhagan, "Problems of Pumpset-Farmers in Tamil Nadu," 341–42.

40. "Electricity in Calcutta: A Promising Enterprise," *Times of India*, April 28, 1899.

41. Particularly in the decade before independence, the ownership status, tariff schedule, and annual profits of the Calcutta Electric Supply Company (CESC) were subject to continuous debate and criticism. Nationalists, especially those in the municipal body, the Calcutta Corporation, repeatedly demanded and occasionally threatened to take over CESC as a public municipal utility, plans that were rejected by the government of Bengal. Proponents of nationalization argued that the firm's profits were plowed into dividends rather than used to lower tariffs for consumers. In the contest between Indian consumers and British shareholders, the former consistently lost. When the decision to nationalize was put to a vote in the corporation in 1937, the vote was split along national lines; European and nominated members voted against nationalization and all elected Indian members voted in favor. For a sense of this debate as it played out in press reports, see "Electricity Charge in Calcutta: Committee to Enquire," *Times of India*, October 23, 1935; "Indian Public Utility Companies: Relations with Local Authorities," *Times of India*, May 20, 1936; "Electricity Supply in Calcutta: Municipalisation Move," *Times of India*, October 6, 1937; "Corporation Motion Not Sanctioned: Calcutta Electric Purchase," *Times of India*, February 1, 1938; "Development of Cheap Electricity," *Times of India*, April 11, 1940.

42. Saha, "River-Basin Planning," 279 and 285.

43. State Statistical Bureau, *Statistical Abstract West Bengal, 1961* (Alipore: West Bengal Government Press, 1965), 521.

44. Hanson, *Process of Planning*, 320. Along with the per capita spending on irrigation and power, Hanson also gives spending on transport and communication, as well as education and culture, and then combines all of these to consider total infrastructural expenditure. Given the size of West Bengal's economy, the comparison is even more stark when these per capita expenditures are considered as a percentage of per capita income. West Bengal's per capita infrastructure spending as a percentage of per capita income across all of these sectors taken together is the lowest in India during these early plan periods.

45. "CPI Delegation Sees Dr. Roy: Power Crisis in Calcutta," *Times of India*, May 6, 1961.

46. Khasnabis, "Economy of West Bengal," 105.

47. See ibid., 104, for data from 2000 on the size of landholdings.

48. Bureau of Applied Economics and Statistics, *Statistical Abstract West Bengal, 1978–1989 (Combined)* (Government of West Bengal, n.d.), 438–46.

49. There is also good reason to be skeptical of the central government's estimate of potential irrigation pumps, given that robust electrification programs have also led to disastrously low water tables in other parts of India. Data for all states are given in Chapter 3.

50. See, especially, Aditi Mukherji, "Political Ecology of Groundwater," 397–401.

51. Indrani Dutta, "Bengal Initiates Steps to Reform Power Sector," *Business Line*, June 8, 2000; Indronil Roychowdhury, "Mamata Could Inherit a Healthy Power Sector," *Financial Express*, May 12, 2011.

52. Jenkins, *Democratic Politics and Economic Reform*, 191.

53. To make this argument, I rely substantially on Kohli's account of the pro-poor policies of the Communist Party of India (Marxist), or CPI(M): Kohli, *State and Poverty in India*, 96–101.

54. For a sample of the media coverage of plant breakdowns and supply restrictions, see: "W. Bengal Power Crisis Deepens," *Times of India*, May 9, 1980; "Power Breakdown Hits Calcutta," *Times of India*, January 21, 1985; "Where Normal Power Is an Event to Rejoice," *Times of India*, September 20, 1989.

55. In 1996, almost twenty years after the CPI(M) came to power, agriculturalists made up just 6.1 percent of the West Bengal state legislature. The largest two groups were political activists (26.2 percent) and teachers (35 percent), both of which make sense in terms of the party's organizational strategies described by Kohli. For a complete sociological analysis of the legislature in West Bengal, see Lama-Rewal, "Resilient *Bhadralok*," 384.

56. "An Area of Darkness," *Economist*, Asian ed., August 4, 2012; "India Blackouts Leave 700 Million Without Power," *Guardian*, July 31, 2012; Jim Yardley and Gardiner Harris, "2nd Day of Power Failures Cripples Wide Swath of India," *New York Times*, July 31, 2012; Sunila S. Kale and Sumit Ganguly, "India's Dark Night: The Politics Behind the Power Failure," *Foreign Affairs* online, August 8, 2012, http://www.foreignaffairs.com/articles/137819/sunila-s-kale-and-sumit-ganguly/indias-dark-night (accessed August 8, 2012).

57. Ministry of Power, "Report of the Enquiry Committee on Grid Disturbance in Northern Region on 30th July 2012 and in Northern, Eastern, and North-Eastern Region on 31st July 2012" (New Delhi: Ministry of Power, 2012).

58. Ibid., 34.

59. Ministry of Power, "Report of the Committee on Manpower, Certification and Incentives for System Operation and Ring Fencing Load Dispatch Centres" (New Delhi: Government of India, 2008).

60. Ministry of Power, "Report of the Enquiry Committee on Grid Disturbance," 35.

61. Text of "PM's Independence Day Speech, 2012," http://pmindia.nic.in/speech-details.php?nodeid=1199 (accessed August 22, 2012).

62. International Energy Agency, "World Energy Outlook 2011—Energy for All, Financing Access for the Poor" (Paris: OECD/IEA, 2011), 16, http://www.iea.org/Papers/2011/weo2011_energy_for_all.pdf (accessed August 31, 2012).

Bibliography

Ahmad, Ausaf. "Regional Pattern of Industrial Licensing." *Social Scientist* (Calcutta) 2, no. 9 (April 1974): 3–17.

Ahmed, Waquar. "Neoliberalism, Corporations, and Power: Enron in India." *Annals of the Association of American Geographers* 100, no. 3 (June 2010): 621–39.

Ambedkar, B. R. "Post-War Development of Electric Power in India." Speech to the Reconstruction Policy Committee meeting, New Delhi, October 25, 1943. Reprinted in *Indian Information*, November 15, 1943, New Delhi: Government of India, 280.

———. "Post-War Electric Power Development." Speech to the Policy Committee on Public Works and Electric Power, New Delhi, February 2, 1945. Reprinted in *Indian Information*, February 15, 1945, New Delhi: Government of India.

———. *Thoughts on Linguistic States*. Aurangabad, 1955.

Ambirajan, S. "State Government Subsidies: The Case of Tamil Nadu." *Economic and Political Weekly* 34, no. 14 (April 1999): 811–21.

Amin, Shahid. *Sugarcane and Sugar in Gorakhpur: An Inquiry into Peasant Production for Capitalist Enterprise in Colonial India*. Delhi: Oxford University Press, 1984.

Amnesty International. "India: The 'Enron Project' in Maharashtra—Protests Suppressed in the Name of Development." New York: Amnesty International, 1997.

Anderson, Benedict. *Imagined Communities: Reflections on the Origin and Spread of Nationalism*. London: Verso, 1983.

Appel, Hilary. "The Ideological Determinants of Liberal Economic Reform: The Case of Privatization." *World Politics* 52, no. 4 (July 2000): 520–49.

Appleby, Paul H. *Public Administration in India: Report of a Survey*. New Delhi: Cabinet Secretariat, 1953.

Attwood, Donald W. "Peasants Versus Capitalists in the Indian Sugar Industry: The Impact of the Irrigation Frontier." *Journal of Asian Studies* 45, no. 1 (November 1985): 59–80.

———. *Raising Cane: The Political Economy of Sugar in Western India*. Boulder, CO: Westview Press, 1992.

Austin, Granville. *The Indian Constitution: Cornerstone of a Nation*. London: Oxford University Press, 1966.

Azad, Nirmal. "Recent Farmers' Agitations in Punjab." *Economic and Political Weekly* 10, no. 17 (April 1975): 702–6.

Bailey, F. G. *Politics and Social Change: Orissa in 1959*. Berkeley: University of California Press, 1963.

Banerjee, Abhijit, and Lakshmi Iyer. "History, Institutions, and Economic Performance: The Legacy of Colonial Land Tenure Systems in India." *American Economic Review* 95, no. 4 (September 2005): 1190–213.

Banerjee, Abhijit, and Rohini Somanathan. "The Political Economy of Public Goods: Some Evidence from India." *Journal of Development Economics* 82, no. 2 (March 2007): 287–314.

Banerjee, Debdas, and Anjan Ghosh. "Indian Planning and Regional Disparity in Growth." In *Economy, Society, and Polity: Essays in the Political Economy of Indian Planning*, edited by Amiya Kumar Bagchi, 104–65. Calcutta: Oxford University Press, 1988.

Banerjee, Kishalay. *Regional Political Parties in India*. Delhi: B.R. Publishing, 1984.

Bardhan, Pranab. *The Political Economy of Development in India*. Oxford: Basic Blackwell, 1984.

Barik, Radhakanta. "Gopabandhu and the National Movement in Orissa." *Social Scientist* (New Delhi) 6, no. 7 (May 1978): 40–52.

Baru, Sanjaya. "Economic Policy and the Development of Capitalism in India: The Role of Regional Capitalists and Political Parties." In *Transforming India: Social and Political Dynamics of Democracy*, edited by Francine Frankel et al., 207–30. New Delhi: Oxford University Press, 2000.

———. *The Political Economy of Indian Sugar: State Intervention and Structural Change*. Delhi: Oxford University Press, 1990.

Bates, Robert. *Markets and States in Tropical Africa: The Political Basis of Agricultural Policies*. Berkeley: University of California Press, 1981.

Bednar, Jenna. *The Robust Federation: Principles of Design*. New York and Cambridge: Cambridge University Press, 2009.

Bentall, Jim, and Stuart Corbridge. "Urban-Rural Relations, Demand Politics and the 'New Agrarianism' in Northwest India: The Bharatiya Kisan Union." *Transactions of the Institute of British Geographers* 21, no. 1 (1996): 27–48.

Bhagwati, Jagdish. *India in Transition: Freeing the Economy*. Oxford: Clarendon Press, 1993.

Bhanage, B. S. "Regulation and Economics of the Electricity Supply Industry in Maharashtra: The Study of a Public Utility." PhD diss., University of Poona, 1976.

Biswas, Rongili, and Sugata Marjit. "Political Lobbying and Fiscal Federalism: Case of Industrial Licenses and Letters of Intent." *Economic and Political Weekly* 37, no. 8 (February 2002): 716–25.

Bombay Electric Supply and Transport Company. "BEST Undertaking: The Brihanmumbai Electric Supply & Transport Undertaking." Company history published by the BEST utility offices. http://www.bestundertaking.com/history.asp (accessed May 2, 2012).

Boone, Catherine. *Political Topographies of the African State: Territorial Authority and Institutional Choice* (New York and Cambridge: Cambridge University Press, 2003).

Brass, Paul. *The Politics of India Since Independence.* 2nd ed. Cambridge: Cambridge University Press, 1994.

Bussell, Jennifer. *Corruption and Reform in India: Public Services in the Digital Age.* New York: Cambridge University Press, 2012.

Byres, Terrence. "Of Neo-populist Pipe-dreams: Daedalus in the Third World and the Myth of Urban Bias." *Journal of Peasant Studies* 6, no. 2 (1979): 210–44.

Carras, Mary C. *The Dynamics of Indian Political Factions: A Study of District Councils in the State of Maharashtra.* London and New York: Cambridge University Press, 1972.

Carter, Anthony T. *Elite Politics in Rural India: Political Stratification and Political Alliances in Western Maharashtra.* London and New York: Cambridge University Press, 1974.

Centre for Monitoring the Indian Economy. *Basic Statistics Relating to the Indian Economy, Vol. 2: States.* Mumbai: CMIE, 1990 and 1991.

———. *Current Energy Scene in India.* Mumbai: CMIE, May 1983.

———. *Profiles of States.* Mumbai: CMIE, March 1997.

Chadaga, Julia Bekman. "Light in Captivity: Spectacular Glass and Soviet Power in the 1920s and 1930s." *Slavic Review* 66, no. 1 (Spring 2007): 82–105.

Chadha, G. K. *The State and Rural Economic Transformation: The Case of Punjab, 1950–1985.* New Delhi: Sage, 1986.

Chakravorty, Sanjoy. "Industrial Location in Post-reform India: Patterns of Interregional Divergence and Intra-regional Convergence." *Journal of Development Studies* 40, no. 2 (December 2003): 120–52.

Chandra, Kanchan. *Why Ethnic Parties Succeed: Patronage and Ethnic Head Counts in India.* Cambridge: Cambridge University Press, 2007.

Chari, Sharad. *Fraternal Capital: Peasant-Workers, Self-Made Men, and Globalization in Provincial India.* Stanford, CA: Stanford University Press, 2004.

Chatterjee, Partha. *A Possible India: Essays in Political Criticism.* New Delhi: Oxford University Press, 1998.

Chattopadhyay, Raghabendra. "The Idea of Planning in India, 1930–1951." PhD diss., Australian National University, 1985.

Chaudhry, Praveen K., Vijay L. Kelkar, and Vikash Yadav. "The Evolution of 'Homegrown Conditionality' in India: IMF Relations." *Journal of Development Studies* 40, no. 6 (August 2004): 59–81.

Chibber, Vivek. *Locked in Place: State-Building and Late Industrialization in India.* Princeton, NJ: Princeton University Press, 2006.

Chousalkar, Ashok S. "Co-operatives: How Has Their Leadership Affected Maharashtra Politics?" In *Politics in Maharashtra,* edited by Usha Thakkar and Mangesh Kulkarni. Bombay: Himalaya, 1995.

Coopersmith, Jonathan. *The Electrification of Russia, 1880–1926.* Ithaca, NY: Cornell University Press, 1992.

Corbridge, Stuart, and John Harriss. *Reinventing India: Liberalization, Hindu Nationalism and Popular Democracy.* Cambridge: Polity Press, 2000.

Das, D. C. "Biju Patnaik: The Vision of a Patriot." *Orissa Review* (February/March 2005): 54–58.

Das, Keshab. "Industrialisation in Orissa: Structure, Policy and Prospects." In *Orissa Development Report,* 200–222. New Delhi: Government of India, Planning Commission, 2001.

Dasgupta, Jyotirindra. "India's Federal Design and Multicultural National Construction." In *The Success of India's Democracy*, edited by Atul Kohli. Cambridge: Cambridge University Press, 2001.

Davis, Diane E. *Discipline and Development: Middle Classes and Prosperity in East Asia and Latin America*. Cambridge: Cambridge University Press, 2004.

Desai, Manali. *State Formation and Radical Democracy in India*. London and New York: Routledge, 2007.

Deshpande, Prachi. "Caste as Maratha: Social Categories, Colonial Policy, and Identity in Early Twentieth-Century Maharashtra." *Indian Economic and Social History Review* 41, no. 1 (2004): 7–32.

D'Souza, Rohan. *Drowned and Damned: Colonial Capitalism and Flood Control in Eastern India*. New Delhi: Oxford University Press, 2006.

Dubash, Navroz K., "The Electricity-Groundwater Conundrum: Case for a Political Solution to a Political Problem." *Economic and Political Weekly* 42, no. 52 (December 2007): 45–55.

———., ed. *Power Politics: Equity and Environment in Electricity Reform*. Washington DC: World Resources Institute, 2002.

———. *Tubewell Capitalism: Groundwater Development and Agrarian Change in Gujarat*. New York: Oxford University Press, 2002.

Dubash, Navroz K., and Sudhir Chella Rajan. "Power Politics: Process of Power Sector Reform in India." *Economic and Political Weekly* 36, no. 35 (September 2001): 3367–90.

Dugger, Celia W. "Surprise in Indian State: Politics of Promising Fails." *New York Times*, October 9, 1999.

Durgaiah, D. *Farmers Movements in India: A Study of Andhra Pradesh*. New Delhi: Classical, 2000.

EPW Research Foundation. *District-Wise Agricultural Database for Maharashtra: 1960–61 to 1997–98*. Mumbai: EPW, 2004.

Erdman, Howard L. "India's Swatantra Party." *Pacific Affairs* 36, no. 4 (Winter 1963–64): 394–410.

Franda, Marcus F. *West Bengal and the Federalizing Process in India*. Princeton, NJ: Princeton University Press, 1968.

Frankel, Francine R. *India's Political Economy, 1947–1977: The Gradual Revolution*. Princeton NJ: Princeton University Press, 1978.

Gandhi, Mohandas K. "Economic Development and Moral Development (1916)." In *Gandhi: 'Hind Swaraj' and Other Writings*, 2nd ed., edited by Anthony J. Parel, 153–61. Cambridge: Cambridge University Press, 2009.

———. "Mass Production Versus Production of the Masses." *Harijan*, February 11, 1934.

Godbole, Madhav. *Unfinished Innings: Recollections and Reflections of a Civil Servant*. New Delhi: Orient Longman, 1996.

Goswami, Manu. *Producing India: From Colonial Economy to National Space*. Chicago: University of Chicago Press, 2004.

Gowda, Chandan. "'Advance Mysore!': The Cultural Logic of a Developmental State." *Economic and Political Weekly* 45, no. 29 (July 2010): 88–95.

Guha, Ramachandra. *India After Gandhi: The History of the World's Largest Democracy*. London: Picador, 2008.

Gulati, Ashok, and Sudha Narayanan. *The Subsidy Syndrome in Indian Agriculture.* New Delhi: Oxford University Press, 2003.

Gupta, M. P. "The Punjab State Electricity Board—An Overview of the Past 50 Years and the Programme of the Ninth Five Year Plan," *Indian Journal of Power and River Valley Development* 46, no. 12 (1996): 261–66.

Haldea, Gajendra. "Crisis of Credibility: Law Must Regulate the Regulator." *Times of India*, January 22, 2004.

———. "Empower Consumers: Plug Loopholes in the Electricity Bill." *Times of India*, April 16, 2003.

Hanson, A. H. *The Process of Planning: A Study of India's Five-Year Plans, 1950–1964.* London: Oxford University Press, 1966.

Harrison, Selig. "Caste and the Andhra Communists." *American Political Science Review* 50, no. 2 (June 1956): 378–404.

Harriss, John. "Comparing Political Regimes Across Indian States: A Preliminary Essay." *Economic and Political Weekly* 34, no. 48 (November 1999): 3367–77.

Harshe, Rajen, and C. Srinivas. "Vote for Development: How Sustainable?" *Economic and Political Weekly* 33, no. 44 (October 1999): 3103–105.

Haya, Barbara, Malini Ranganathan, and Sujit Kirpekar. "Barriers to Sugar Mill Cogeneration in India: Insights into the Structure of Post-2012 Climate Financing Instruments." *Climate and Development* 1, no. 1 (2009): 66–81.

Headrick, Daniel. *The Tools of Empire: Technology and European Imperialism in the Nineteenth Century.* New York: Oxford University Press, 1981.

Henderson, P. D. *India: The Energy Sector.* Washington, DC: World Bank, 1975.

Herring, Ronald. *Land to the Tiller: The Political Economy of Agrarian Reform in South Asia.* New Haven, CT: Yale University Press, 1983.

Herring, Ronald, and Rina Agarwala. "Introduction: Restoring Agency to Class: Puzzles from the Subcontinent." *Critical Asian Studies* 38, no. 4 (December 2006): 323–56.

Hughes, Thomas. *Networks of Power: Electrification in Western Society, 1880–1930.* Baltimore, MD: Johns Hopkins University Press, 1983.

Human Rights Watch. *The Enron Corporation: Corporate Complicity in Human Rights Violations.* New York: HRW, 1999.

Inglis, C. C., and V. K. Gokhale. "Note on the Outturn and Profit from Sugarcane Crops Grown in the Deccan Canal Tracts." Bombay Public Works Department, Technical Paper No. 21. Bombay: Government Central Press, 1928.

International Energy Agency. *World Energy Outlook 2011—Energy for All, Financing Access for the Poor.* Paris: OECD/IEA, 2011.

Jaffrelot, Christophe. *Saffron Modernity in India: Narendra Modi and His Experiment with Gujarat.* London: Hurst, forthcoming 2014.

Jain, Varinder. "Political Economy of the Electricity Subsidy: Evidence from Punjab." *Economic and Political Weekly* 41, no. 38 (September 2006): 4072–80.

Jalal, Ayesha. *Democracy and Authoritarianism in South Asia: A Comparative and Historical Perspective.* Cambridge: Cambridge University Press, 1995.

Jalan, Bimal. *India's Economic Crisis: The Way Ahead.* New Delhi: Oxford University Press, 1991.

Jena, B. B. "Political Parties in Orissa." In *State Politics in India*, edited by Iqbal Narain, D. B. Mathur, and Sushil Kumar, 485–502. Meerut: Meenakshi Prakashan, 1967.

Jenkins, Rob. *Democratic Politics and Economic Reform*. Cambridge and New York: Cambridge University Press, 1999.

Jodhka, Surinder S. "Nation and Village: Images of Rural India in Gandhi, Nehru and Ambedkar." *Economic and Political Weekly* 37, no. 32 (August 2002): 3343–53.

Johri, S. N. "Nalco's Role in Orissa's Development." In *Reference Orissa: An Indian State of Eastern Region*, 342–46. New Delhi: Enterprising Publishers, 2000.

Joseph, Kelli Lynn. "The Politics of Power: Electricity Reform in India." PhD diss., University of Virginia, 2008.

Kamat, G. S. *Management of Co-operative Sugar Factories in Maharashtra*. Bombay: Maharashtra Rajya Sahakari Sangh, 1976.

Kannan, K. P., and N. Vijayamohanan Pillai. *Plight of the Power Sector in India: Inefficiency, Reform, and Political Economy*. Thiruvananthapuram, Kerala: Centre for Development Studies, 2002.

Kanungo, Pabak. "Biju Patnaik: A Profile." In *Legendary Biju: The Man and the Mission*, edited by K. P. Mohanty, 59–75. New Delhi: Kaveri, 2002.

Kaviraj, Sudipta. "A Critique of the Passive Revolution." *Economic and Political Weekly* 23, nos. 45–47 (November 1988): 2429–44.

Khagram, Sanjeev. *Dams and Development: Transnational Struggles for Water and Power*. Ithaca, NY, and London: Cornell University Press, 2004.

Khasnabis, Ratan. "The Economy of West Bengal." *Economic and Political Weekly* 43, no. 52 (December 2008): 103–15.

Khekale, Narayan R. *The Pressure Politics in Maharashtra, India*. Mumbai: Himalaya, 1999.

Khilnani, Sunil. *The Idea of India*. New York: Farrar, Straus and Giroux, 1998.

Khosla, A. N. *Orissa's Decade of Destiny: 1963–1973, A Master Plan for the Integrated Development of the River Basins of Orissa*. Madhupatna, Cuttack: Government of Orissa, 1963.

King, John A. *Economic Development Projects and Their Appraisal: Cases and Principles from the Experience of the World Bank*. Baltimore, MD: Johns Hopkins University Press, 1967.

Kirk, Jason. *India and the World Bank: The Politics of Aid and Influence*. London and New York: Anthem Press, 2010.

Klingensmith, Daniel. *"One Valley and a Thousand": Dams, Nationalism, and Development*. New Delhi: Oxford University Press, 2007.

Kobo, Kensuke. "Cropping Pattern Changes in Andhra Pradesh During the 1990s: Implications for Micro-Level Studies." In *Agricultural Production, Household Behavior, and Child Labor in Andhra Pradesh*, edited by Seiro Ito. Wakaba, Japan: Institute for Developing Economies, 2005.

Kohli, Atul. "The NTR Phenomenon in Andhra Pradesh: Political Change in a South Indian State." *Asian Survey* 28, no. 10 (October 1988): 991–1017.

———. *Poverty amid Plenty in the New India*. Cambridge and New York: Cambridge University Press, 2012.

———. *The State and Poverty in India: The Politics of Reform*. Cambridge and New York: Cambridge University Press, 1987.

Kumar, Ashutosh, and T. R. Sharma. "Legislative Elite in Punjab: A Socio-political Study." In *Rise of the Plebeians? The Changing Face of Indian Legislative Assemblies*, edited by Christophe Jaffrelot and Sanjay Kumar, 151–63. New Delhi: Routledge, 2009.

Kumar, Sanjay. "Janata Regionalized: Contrasting Bases of Electoral Support in Bihar and Orissa." In *Regional Reflections: Comparing Politics Across India's States*, edited by Rob Jenkins, 111–38. New Delhi: Oxford, 2004.

Kurien, C. T. "Dynamics of Rural Transformation: A Case Study of Tamil Nadu." *Economic and Political Weekly* 15, nos. 5–7 (February 1980): 365–90.

Lama-Rewal, Stephanie Tawa. "The Resilient *Bhadralok*: A Profile of the West Bengal MLAs." In *Rise of the Plebians? The Changing Face of Indian Legislative Assemblies*, edited by Christophe Jaffrelot and Sanjay Kumar, 361–92. London: Routledge, 2009.

Lele, Jayant. *Elite Pluralism and Class Rule: Political Development in Maharashtra, India.* Toronto: University of Toronto Press, 1981.

Lindberg, Staffan. "New Farmers' Movements in India as Structural Response and Collective Identity Formation: The Cases of the Shetkari Sanghatana and the BKU." *Journal of Peasant Studies* 21, nos. 3–4 (April/July 1994): 95–125.

Lipton, Michael. *Why Poor People Stay Poor: Urban Bias in World Development.* Cambridge, MA: Harvard University Press, 1977.

Ludden, David. "India's Development Regime." In *Colonialism and Culture*, edited by Nicholas B. Dirks, 247–87. Ann Arbor: University of Michigan Press, 1992.

Madrid, Raul. *Retiring the State: The Politics of Pension Privatization in Latin America and Beyond.* Stanford, CA: Stanford University Press, 2003.

Maharashtra Economic Development Council. *Power Development in Maharashtra.* Bombay: MEDC, 1975.

Manibog, Fernando, Rafael Dominguez, and Stephan Wegner. *Power for Development: A Review of the World Bank Group's Experience with Private Participation in the Electricity Sector.* Washington, DC: World Bank, 2003.

Mann, Harold, and N. V. Kanitkar. *Land and Labour in a Deccan Village: Study No. 2* (Bombay: Oxford University Press, 1921).

Mann, Michael. "The Autonomous Power of the State: Its Origins, Mechanisms, and Results." *Archives of European Sociology* 25 (1984): 185–213.

———. "Infrastructural Power Revisited." *Studies in Comparative International Development* 43, nos. 3–4 (2008): 355–65.

Manor, James. "Explaining Political Trajectories in Andhra Pradesh and Karnataka." In *Regional Reflections: Comparing Politics Across India's States*, edited by Rob Jenkins, 255–84. New Delhi: Oxford, 2004.

———. "Where Congress Survived: Five States in the Indian General Election of 1977." *Asian Survey* 18, no. 8 (August 1978): 785–803.

Meher, Rajkishor. "Infrastructure." In *Orissa Development Report*, 54–70. New Delhi: Government of India, Planning Commission, 2001.

Mehta, Abhay. *Power Play: A Study of the Enron Project.* Mumbai: Orient Longman, 1999.

Menon, Meena. "Nation in a State: That Maratha Demand, Again." *Hindu*, June 12, 2012.

Menon, Nivedita, and Aditya Nigam. *Power and Contestation: India Since 1989.* Hyderabad: Orient Longman, 2008.

Migdal, Joel S., Atul Kohli, and Vivienne Shue. *State Power and Social Forces: Domination and Transformation in the Third World.* Cambridge and New York: Cambridge University Press, 1994.

Min, Brian. "Distributing Power: Electrifying the Poor in India." Working paper, October 2010.

Mishra, Banikanta. "Agriculture, Industry and Mining in Orissa in the Post-Liberalisation Era: An Inter-district and Inter-state Panel Analysis." *Economic and Political Weekly* 45, no. 20 (May 2010): 49–68.

Mishra, Digambar. *Political Behavior of Indian State Governors: A Study of the Role of Governor in Orissa*. Delhi: Kalinga, 2002.

Misra, Surya Narayan. *Party Politics and Electoral Choice in an Indian State*. Delhi: Ajanta, 1989.

Mitchell, Timothy. *Carbon Democracy: Political Power in the Age of Oil*. London: Verso, 2011.

———. *Rule of Experts: Egypt, Techno-politics, Modernity*. Berkeley: University of California Press, 2002.

Mitra, Ashok K., and S. W. Muranjan. *Economic Benefits of Rural Electrification in Maharashtra—A Study of Four Districts*. Pune: Gokhale Institute of Politics and Economics, 1980.

Mohanty, B. K. "Mineral Development Prospects." In *Reference Orissa: An Indian State of Eastern Region*, 453–58. New Delhi: Enterprising Publishers, 2000.

Mohanty, Balaram. *Orissa: A District-Wise Analysis*. Bhubaneswar, Orissa: Vidyaratna Press, 1997.

Mohanty, Manoranjan. "Adrift Middle Class." *Economic and Political Weekly* 21, no. 33 (August 1986): 1445–46.

———. "Class, Caste and Dominance in a Backward State: Orissa." In *Dominance and State Power in Modern India, Vol. 2*, edited by Francine R. Frankel and M. S. A. Rao, 321–66. Delhi and New York: Oxford, 1989.

Mohanty, U. K. "Potential for Coal Mining." In *Reference Orissa: An Indian State of Eastern Region*, 459–66. New Delhi: Enterprising Publishers, 2000.

Mukherjee, Ambarish, and Nithya Subramanian. "Corporates That Have Stood the Test of Time Since Independence." *Hindu Business Line*, January 26, 2003.

Mukherji, Aditi. "Political Ecology of Groundwater: The Contrasting Case of Water-Abundant West Bengal and Water-Scarce Gujarat, India." *Hydrogeology Journal* 14, no. 3 (2006): 392–406.

Mukherji, Rahul. *Globalization and Deregulation: Ideas, Interests and Institutional Change in India*. New Delhi: Oxford University Press, 2014.

———. "India's Aborted Liberalization—1966." *Pacific Affairs* 73, no. 3 (Autumn 2000): 375–92.

Muraleedharan, V. R. "The Electrification of Madras City, 1905–1914." Paper presented to the Third World Economic History and Development Conference, University of Manchester, September 13–15, 1991.

Murillo, Maria Victoria. *Political Competition, Partisanship, and Policy Making in Latin American Public Utilities*. Cambridge and New York: Cambridge University Press, 2009.

Naoroji, Dadabhai. *Poverty and Un-British Rule in India*. London: Swan Sonnenschein, 1901.

National Planning Committee, *Report of the Sub-committee on Power and Fuel*. Bombay: Vora, 1947.

Nayak, Arun Kumar. "Big Dams and Protests in India: A Study of Hirakud Dam." *Economic and Political Weekly* 45, no. 2 (January 2010): 69–73.

Nehru, Jawaharlal. *Glimpses of World History*. 4th ed. Suffolk, UK: Richard Clay, 1949.

Nye, David E. *Electrifying America: Social Meanings of a New Technology: 1880–1940*. Cambridge, MA: MIT Press, 1990.

O'Donnell, Guillermo. "On the State, Democratization and Some Conceptual Problems: A Latin American View with Glances at Some Postcommunist Countries." *World Development* 21, no. 8 (1993): 1355–69.

O'Hanlon, Rosalind. *Caste, Conflict and Ideology: Mahatma Jotirao Phule and Low Caste Protest in Nineteenth-Century Western India*. Cambridge and New York: Cambridge University Press, 1985.

Omvedt, Gail. *Cultural Revolt in a Colonial Society: The non Brahman Movement in Western India, 1873–1930*. Bombay: Scientific Socialist Education Trust, 1976.

Pachauri, R. K. *Energy and Economic Development in India*. New York: Praeger, 1977.

Padhi, Sakti. "Economic Growth, Structural Change and Workforce Characteristics." In *Orissa Development Report*, 71–106. New Delhi: Government of India, Planning Commission, 2001.

Palshikar, Shreeyash. "Breaking Bombay, Making Maharashtra: Media, Identity Politics and State Formation in Modern India." PhD diss., University of Chicago, 2007.

Palshikar, Suhas, Rajeshwari Deshpande, and Nitin Birmal. "Maharashtra Polls: Continuity Amidst Social Volatility." *Economic and Political Weekly* 44, no. 48 (November 2009): 42–47.

Parekh, Bhikhu. "Nehru and the National Philosophy of India." *Economic and Political Weekly* 26, nos. 1–2 (January 1991): 35–48.

Pati, Biswamoy. "Dialectics of Transition: Orissa, 1943–1950." *Economic and Political Weekly* 27, no. 7 (February 1992): 353–64.

———. *Resisting Domination: Peasants, Tribals and the National Movement in Orissa: 1920–1950*. New Delhi: Manohar, 1993.

Patra, S. C. *Formation of the Province of Orissa: The Success of the First Linguistic Movement in India*. Calcutta: Punthi Pustak, 1979.

Phadke, Amol, and Sudhir Chella Rajan. "Electricity Reforms in India: Not Too Late to Go Back to the Drawing Board." *Economic and Political Weekly* 38, no. 29 (July 2003): 3061–72.

Phadke, Y. D. "The Samyukta Maharashtra Movement: 1946–1960." In *Language Movements in India*, edited by E. Annamalai, 118–34. Mysore: Central Indian Institute of Languages, 1979.

Prakash, Gyan. *Another Reason: Science and the Imagination of Modern India*. Princeton, NJ: Princeton University Press, 1999.

Prasad, C. R., K. Krishna Prasad, and A. K. N. Reddy. "Bio-Gas Plants: Prospects, Problems and Tasks." *Economic and Political Weekly* 9, nos. 32–34 (August 1974): 1347–64.

Prayas Energy Group. "A Good Beginning but Challenges Galore." *Prayas Occasional Report* 1, 2003.

Przeworski, Adam. *Democracy and the Market: Political and Economic Reforms in Eastern Europe and Latin America*. Cambridge and New York: Cambridge University Press, 1991.

Purkayastha, Prabir. "The Electricity Bill 2003—From Crisis to Disaster," *People's Organ* (weekly of the Communist Party of India [Marxist]), 27.18, May 4, 2003, http://www.cpim.org/pd/2003/0504/05042003_snd.htm (accessed June 2013).

Purkayastha, Prabir, and Vijay Prashad. *Enron Blowout: Corporate Capitalism and Theft of the Global Commons*. New Delhi: Leftword, 2002.

Rajagopal, A., and P. Anbazhagan. "Problems of Pumpset-Farmers in Tamil Nadu." *Economic and Political Weekly* 24, no. 7 (February 1989): 341–42.

Ramana, D. V., Banikanta Mishra, and Birendra K. Nayak. "Power Sector Reform in Orissa: A Case Study of Restructuring." In *Orissa Development Report*, 376–408. New Delhi: Government of India, Planning Commission, 2001.

Rao, Anupama. *The Caste Question: Dalits and the Politics of Modern India*. Berkeley: University of California Press, 2009.

Rao, Yenda Srinivasa. "Electricity, Politics and Regional Economic Imbalance in Madras Presidency, 1900–1947." *Economic and Political Weekly* 45, no. 23 (June 2010): 59–66.

Rath, Ramakanta. *The Law of Zamindari Abolition*. Cuttack, Orissa: Lark Books, 1965.

Reddy, G. Ram. "The Politics of Accommodation: Caste, Class and Dominance in Andhra Pradesh." In *Dominance and State Power in Modern India, Vol. I*, edited by Francine Frankel and M. S. A. Rao, 265–321. Delhi: Oxford University Press, 1989.

Rosenthal, Donald B. *The Expansive Elite: District Politics and State Policy-Making in India*. Berkeley: University of California Press, 1977.

Roy, Srirupa. *Beyond Belief: India and the Politics of Postcolonial Nationalism*. Durham, NC: Duke University Press, 2007.

Rudolph, Lloyd I., and Susanne Hoeber Rudolph. "Federalism as State Formation in India: A Theory of Shared and Negotiated Sovereignty." *International Political Science Review* 31, no. 5 (2010): 553–72.

———. "The Iconization of Chandrababu: Sharing Sovereignty in India's Federal Market Economy." *Economic and Political Weekly* 36, no. 18 (May 2001): 1541–52.

———. *In Pursuit of Lakshmi: The Political Economy of the Indian State*. Chicago: University of Chicago Press, 1987.

Ruet, Joel, ed. *Against the Current: Organizational Restructuring of State Electricity Boards*. New Delhi: Manohar, 2003.

Rutnagur, S. M., ed. *Electricity in India: Being a History of the Tata Hydro-Electric Project with Notes on the Mill Industry in Bombay and the Progress of Electric Drive in Indian Factories*. Bombay: Indian Textile Journal, 1912.

Saez, Lawrence. *Federalism Without a Centre: The Impact of Political and Economic Reform on India's Federal System*. New Delhi: Sage Publications, 2002.

Saha, Suranjit Kumar, "River-Basin Planning in the Damodar Valley of India." *Geographical Review* 69, no. 3 (July 1979): 273–87.

Samant, D. R., and M. A. Mulky. *Organisation and Finance of Industries in India*. Calcutta: Longmans, Green, 1937.

Samyukta Maharashtra Parishad. *Reorganization of States in India with Particular Reference to the Formation of Maharashtra: Being the Memorandum Submitted by the Samyukta Maharashtra Parishad to the States Reorganization Commission*. Bombay: Samukta Maharashtra Parishad, 1954.

Sant, Girish, and Shantanu Dixit. "Beneficiaries of IPS Subsidy and Impact of Tariff Hike." *Economic and Political Weekly* 31, no. 51 (December 1996): 3315–21.

Sarkar, Jayanta. "Letter from India: Political Balkanization Continues to Poison IPP Environment." *Electrical World* 210, no. 11 (November 1996): 66.

Sawhney, Upinder. "Subnational Reforms and Public Policy Issues in Punjab." *Journal of Punjab Studies* 19, no. 1 (Spring 2012): 49–66.

Schamis, Hector. *Re-forming the State: The Politics of Privatization in Latin America and Europe*. Ann Arbor: University of Michigan Press, 2002.

Schivelbusch, Wolfgang. *Disenchanted Night: The Industrialization of Light in the Nineteenth Century*. Translated from German by Angela Davis. Berkeley: University of California Press, 1995.

Segal, Howard. *Technological Utopianism in American Culture*. Chicago: University of Chicago Press, 1985.

Shah, Alpa. *In the Shadows of the State: Indigenous Politics, Environmentalism, and Insurgency in Jharkhand, India*. Durham, NC: Duke University Press, 2010.

Shinde, A. S. *The Working of the Maharashtra Legislative Assembly*. Aurangabad: Kailash, 2000.

Singh, Baldev, ed. *Jawaharlal Nehru on Science and Society: A Collection of His Writings and Speeches*. New Delhi: Nehru Memorial Museum and Library, 1988.

Singh, Charan. *India's Economic Policy: The Gandhian Blueprint*. New Delhi: Vikas, 1978.

Singh, Nirvikar, and Garima Vasishtha. "Patterns in Centre-State Fiscal Transfers: An Illustrative Analysis." *Economic and Political Weekly* 39, no. 45 (November 2004): 4897–903.

Singh, Prerna. "We-ness and Welfare: A Longitudinal Analysis of Social Development in Kerala, India." *World Development* 39, no. 2 (2011): 282–93.

Sinha, Aseema. "Reforming Public Services in a High Growth State: The Case of Gujarat." In *Public Service Delivery in India: Understanding the Reform Process*, edited by Vikram M. Chand, 126–76. New Delhi: Oxford University Press, 2010.

———. *The Regional Roots of Developmental Politics in India: A Divided Leviathan*. Bloomington: Indiana University Press, 2005.

Sinha, B. N. *Geography of Orissa*. 3rd ed. New Delhi: National Book Trust, 1999.

Sirsikar, V. M. *Politics of Modern Maharashtra*. Bombay: Orient Longman, 1995.

Snyder, Richard. "Scaling Down: The Subnational Comparative Method." *Studies in Comparative International Development* 36, no. 1 (Spring 2001): 93–110.

Soifer, Hillel. "State Infrastructural Power: Approaches to Conceptualization and Measurement." *Studies in Comparative International Development* 43, nos. 3–4 (September 2008): 231–51.

Soifer, Hillel, and Matthias vom Hau, eds. Special issue. *Studies in Comparative International Development* 43, nos. 3–4 (September 2008).

Srinivas, M. N. "The Social Structure of a Mysore Village." In *Village India: Studies in the Little Community*, edited by McKim Marriott. Chicago: University of Chicago Press, 1955.

Subramanian, Narendra. *Ethnicity and Populist Mobilization: Political Parties, Citizens, and Democracy in South India*. New Delhi: Oxford, 1999.

Sur, Mona, and Dina Umali-Deininger. "Public Expenditures and Subsidies in Indian Surface Irrigation: Who Benefits." Paper presented at Water Week 2003, conference organized by World Bank, Washington DC, March 4–6, 2003.

Swades: We, the People. Directed by Ashutosh Gowariker. India: UTV Motion Pictures, 2004.

Taylor, C. *India: Economic Issues in the Power Sector*. Washington, DC: World Bank, South Asia Regional Office, 1979.

Teitelbaum, Emmanuel. *Mobilizing Restraint: Democracy and Industrial Conflict in Post-Reform South Asia*. Ithaca, NY: ILR Press, 2011.

Tendler, Judith. *Electric Power in Brazil: Entrepreneurship in the Public Sector*. Cambridge, MA: Harvard University Press, 1968.

Thillai Rajan, A. "Power Sector Reform in Orissa: An Ex-post Analysis of the Causal Factors." *Energy Policy* 28, no. 10 (2000): 657–69.

Tongia, Rahul. "The Political Economy of Indian Power Sector Reforms." In *The Political Economy of Power Sector Reform: The Experiences of Five Major Developing Countries*, edited by David G. Victor and Thomas C. Heller, 109–74. Cambridge and New York: Cambridge University Press, 2007.

Upadhya, Carol Boyak. "The Farmer-Capitalists of Coastal Andhra Pradesh." *Economic and Political Weekly* 23, no. 27 (July 1988): 1376–82.

———. "The Farmer-Capitalists of Coastal Andhra Pradesh." *Economic and Political Weekly* 23, no. 28 (July 1988): 1433–42.

Varshney, Ashutosh. *Democracy, Development, and the Countryside: Urban-Rural Struggles in India.* Cambridge: Cambridge University Press, 1995.

Venkataraman, K. *Power Development in India: The Financial Aspects.* New Delhi: Wiley Eastern Private, 1972.

Vinaik, M. *Sarvodaya & Electricity.* Madras: Kumarappa, 1959.

Visvesvaraya, M. *Memoirs of My Working Life.* Bombay: Printed by F. Borton; Bangalore: Published by M. Visvesvaraya, 1951.

———. *Planned Economy for India.* 2nd ed. Bangalore: Bangalore Press, 1936.

Vora, Rajendra. "Shift of Power from Rural to Urban Sector." *Economic and Political Weekly* 31, nos. 2–3 (January 1996): 171–73.

Vyasulu, Vinod, and A. V. Arun Kumar. "Industrialisation in Orissa: Trends and Structure." *Economic and Political Weekly* 32, no. 22 (May 1997): M46–M53.

Wade, Robert. "The System of Administrative and Political Corruption: Canal Irrigation in South India." *Journal of Development Studies* 18, no. 3 (1982): 287–328.

Washbrook, David A. *The Emergence of Provincial Politics: The Madras Presidency, 1870–1920.* New Delhi: Vikas, 1977; Cambridge: Cambridge University Press, 1976.

Waterbury, John. *Exposed to Innumerable Delusions: Public Enterprise and State Power in Egypt, India, Mexico, and Turkey.* Cambridge: Cambridge University Press, 1993.

Weyland, Kurt G. *The Politics of Market Reform in Fragile Democracies: Argentina, Brazil, Peru, and Venezuela.* Princeton, NJ: Princeton University Press, 2002.

Williamson, A. V. "Indigenous Irrigation Works in Peninsular India." *Geographical Review* 21, no. 4 (October 1931): 613–26.

World Bank. *Asia Sustainable and Alternative Energy Program (ASTAE) Annual Status Report 18, Fiscal Year 2010.* Washington, DC: World Bank Group, 2010.

World Bank Country Operations, Industry and Finance Division, South Asia Region. "Andhra Pradesh: Agenda for Economic Reforms." New Delhi: World Bank, January 1997.

World Bank Energy and Infrastructure Operations Division, Country Department II, South Asia Region. *Staff Appraisal Report, India: Orissa Power Sector Restructuring Project.* New Delhi: World Bank, 1996.

World Bank Energy and Infrastructure Unit, South Asia Region. "Implementation Completion Report on a Loan in the Amount of US$210 Million to the Government of India for Andhra Pradesh Power Sector Restructuring Project." Washington, DC: World Bank, February 20, 2004.

World Bank Energy Sector Unit, South Asia Region. "Andhra Pradesh Power Sector Restructuring Program." Report No. 18849IN. New Delhi: World Bank, January 1999.

World Bank Operations and Evaluation Department. "India: The Challenges of Development, Overview of Sectoral Assistance Evaluations." OED India CAE Working Paper Series No. 16. Washington, DC: World Bank, 2002.

———. "Reforming India's Energy Sector (1978–99)." *Precis* 206 (Spring 2001).

Ziblatt, Daniel. "Why Some Cities Provide More Public Goods Than Others: A Subnational Comparison of the Provision of Public Goods in German Cities in 1912."

Studies in Comparative International Development 43, nos. 3–4 (September 2008): 273–89.

Colonial-Era Government Publications

Census of India, 1921. Volume VII: Bihar and Orissa, Part I. Patna: Government Printing, Bihar and Orissa, 1923.
Census of India, 1931. Volume VII: Bihar and Orissa, Part I. Patna: Government Printing, Bihar and Orissa, 1933.
Indian Information.
Indian Information Series.
Office of the Electrical Commissioner. *Public Electricity Supply: All India Statistics, 1944.* Delhi: Government of India, 1947.
Report of the Madras Electricity Board Committee, 1935–36, Volume II. Madras: Government Press, 1936.

Government of Andhra Pradesh Publications

Andhra Pradesh Department of Industries and Commerce. *Industrial Opportunities in Andhra Pradesh.*
Andhra Pradesh Electricity Reform Act, 1998.
Andhra Pradesh State Electricity Board. *Administration Report*, various years.
Bureau of Economics and Statistics. *Five Decades of Andhra Pradesh.*
Finance and Planning Department. *Sixth Five-Year Plan, 1980–1985, Draft Vol. I.*
Planning Department. *Economic Development of Andhra Pradesh, 1951–1968.*
Transmission Corporation of Andhra Pradesh, *Power Development in Andhra Pradesh (Statistics) 1999–2000.*

Government of Maharashtra Publications

Finance Department (Planning), *Draft Third Five-Year Plan of the Maharashtra State: 1961–1962 to 1965–66*, 1960.
———. *Third Five-Year Plan Maharashtra State: Annual Plan for 1965–66, Part I-Outline*, 1965.
Human Development Report Maharashtra 2002.
Maharashtra Legislative Assembly. Committee on Public Undertakings Reports, various years.
Maharashtra State Electricity Board, *Administrative Report*, various years.
"Report of the Energy Review Committee, Part I" (*Godbole Committee Report*).

Government of Odisha Publications

Irrigation and Power Department, Orissa, *Report on Tikerpara Dam Project*, 1964.
Orissa (India), and H. R. Khanna. *Report of the Commission of Inquiry*, 1969.
Orissa Electricity Reform Act, 1995.
Orissa State Electricity Board. "About Us," 1976.
Planning Department (Orissa). "Third Five-Year Plan Orissa, Annual Plan 1965–66."

Report of the Committee on Power Sector Reform of Orissa, 2001 (Kanungo Committee Report).

State Evaluation Organisation: Planning and Coordination Department. *Report on the Evaluation of Rural Electrification Programme in Orissa*, 1982.

Indian Central Government Publications

Central Electricity Authority. *Electric Power Survey of India*, various years.

———. *Power Survey of India*, various years.

———. *Report of the Committee to Recommend Next Higher Size of Coal Fired Thermal Power Stations*. New Delhi: Government of India, Ministry of Power, November 2003.

Central Water and Power Commission. *Public Electricity Supply: All India Statistics*, various years.

Constituent Assembly Debates, various volumes.

Election Commission of India. "Statistical Report on the General Elections to the Legislative Assembly of Orissa," various years.

Electricity (Supply) Act, 1948.

Gazette of India.

Lok Sabha Debates, various volumes.

Ministry of Agriculture. *Report of the Sugar Industry Enquiry Commission: 1974, Volumes I and II*, 1974.

Ministry of Irrigation and Power. *Power Economy Committee, Report of Study Group, Vol. I*, 1973.

———. *Report of the Power Economy Committee*, 1971.

Ministry of Power. *Annual Report*, various years.

———. "Report of the Committee on Manpower, Certification and Incentives for System Operation and Ring Fencing Load Dispatch Centres," 2008.

———. "Report of the Enquiry Committee on Grid Disturbance in Northern Region on 30th July 2012 and in Northern, Eastern, and North-Eastern Region on 31st July 2012," 2012.

Planning Commission. "Annual Report on the Working of State Electricity Boards and Electricity Departments." New Delhi: Government of India, various years.

———. *Indian Planning Experience: A Statistical Profile*, 2001.

———. *Tenth Five-Year Plan (2002–2007), Volume 3*.

Programme Evaluation Organisation, Planning Commission. *Report on the Evaluation of the Rural Electrification Programme*, 1965.

Report of the Power Tariff Policy Committee (Venkataraman Committee Report). New Delhi: Government of India, 1964.

Report of the Select Committee on the Electricity (Supply) Bill, 1947.

Reserve Bank of India. *Report of the All-India Rural Credit Review Committee*, 1969.

Other State Government Publications

Bureau of Applied Economics and Statistics. *Statistical Abstract West Bengal, 1978–1989 (Combined)*. Government of West Bengal.

Madras State Electricity Board. Administration Report, 1961–1962.

Planning Department, *Third Five-Year Plan, Madras State: Draft Outline*. Madras: Government of Madras, 1960.

State Statistical Bureau. *Statistical Abstract West Bengal, 1961*. Alipore: West Bengal Government Press, 1965.

South Asian and International News Sources

Business Line (India)
Economic Times (India)
Economist
Financial Express (India)
Financial Times (India)
FT Energy Newsletters—Global Private Power
FT Energy Newsletters—International Coal Report
FT Energy Newsletters—Power Asia
Global Power Report
Guardian (UK)
Hindu (India)
India Today
Outlook (India)
Statesman (India)
Times of India
Tribune (India)

Index

Page numbers followed by *t* and *map* denote entries for tables and maps.